BOB MILLER'S SAT® MATH FOR THE CLUELESS

SAT® MATH

OTHER TITLES IN BOB MILLER'S CLUELESS SERIES

Bob Miller's Algebra for the Clueless
Bob Miller's Calc for the Clueless: Precalc with Trig, Second Edition
Bob Miller's Calc for the Clueless: Calc I, Second Edition
Bob Miller's Calc for the Clueless: Calc II, Second Edition
Bob Miller's Calc for the Clueless: Calc III

SAT® MATH

Robert Miller

Mathematics Department
City College of New York

McGraw-Hill

New York   San Francisco   Washington, D.C.   Auckland   Bogotá
Caracas   Lisbon   London   Madrid   Mexico City   Milan
Montreal   New Delhi   San Juan   Singapore
Sydney   Tokyo   Toronto

To my wife, Marlene, I dedicate this book and everything else
I ever do to you. I love you very, very much.

BOB MILLER'S SAT MATH FOR THE CLUELESS

    4 5 6 7 8 9 10 11 12 13 14 15 16 17 18 19 20 DOC DOC   0 2

ISBN 0-07-043432-8

SAT® is a trademark of the Educational Testing Service, which does not endorse
this book.

Sponsoring Editor: Barbara Gilson
Production Supervisor: Sherri Souffrance
Editing Supervisor: Maureen Walker
Project Supervision: North Market Street Graphics
Photo: Eric Miller

McGraw-Hill

A Division of The McGraw·Hill Companies

CONTENTS

TO THE STUDENT

This book is written for you: not your teacher, not your next door neighbor, not for anyone but you. It is written so that you might improve your math SAT score by 50 or 100 points, or maybe even a little more.

However, as much as I hate to admit it, I am not perfect. If you find anything that is unclear or should be added to the book, please let me know. If I can help you get a better score, please write me c/o McGraw-Hill, Schaum Division, Two Penn Plaza, New York, New York 10121-2298. Please enclose a self-addressed stamped envelope. If you want the response before your actual SAT, please write at least 30 days before your test. If you need more help on the basics, my *Algebra for the Clueless,* published around the same time as this book (it also has some arithmetic), will help. For advanced stuff, the *Calc for the Clueless* series will help: Precalc with Trig, Calc I, Calc II, and Calc III.

Now, enjoy the book and learn!!!!!

HOW TO USE THIS BOOK

This book, if used properly, is designed for you to do great on your SAT. It is written in small bits with immediate problems. Read carefully the problems that explain each skill, and try each of the problems that are there for you to practice. Buuuut. . . . beware!!! Even the best students, the first or second time through, make mistakes, lots of mistakes. This is NOT important. The only day that counts is the day you take the SAT. Next, read the solution to each problem. Make sure you understand what was done in each problem. Quality study is much more important than quantity. Speed is NOT important until you get to the actual test. You will automatically go faster when you know the material and know the tricks. Give yourself enough time before the SAT so that you can learn everything. Enjoy the practice. I love to do these kinds of questions. I hope you soon will too.

I WANT YOU TO IMPROVE 50 POINTS OR 100 POINTS OR MORE ON THE MATH SAT®

You: What is the math SAT?

Me: It is a game!

You: What is the object of the game??

Me: What is the object of any game?

You: To win.

Me: Right. And what do you win?

You: Well, what DO I win??

Me: The college of your choice!! Let me give you an example of the game. Remember, you should do no writing with this problem, none at all.

You: But the teacher always told me to show all the work and write everything down. I can't do it any other way. I can't. I can't! I CAN'T!!!!!!

Me: Sure you can, but it may take a little time to learn. Remember, it is not important that you get the problem right the first or second time. The only important time is the day of the SAT.

You: OK, OK. Llllet's sssseee the problem.

Me: If $2x - 1 = 80$, what is $2x - 3$? No, no, no! Don't solve for x. Just look at $2x - 1$ . . . then look at $2x - 3$. . . .

You: It looks like two less.

Me: That's right!!! So the answer is . . .

You: 78!!! 78!!!!

Me: That's terrific. 2x − 1 to 2x − 3 means you go down 2. 80 − 2 = 78.

You: I would never think of this on my own.

Me: Maybe not now, but after you read this book, you'll be muuuuch better!! It will help you in your regular math class also. It will also make math more fun because you'll be able to work quicker with less writing.

You: I'll bet you have to know a zillion facts to improve 50 or 100 points on the math SAT.

Me: Oddly, this is not true. You need to know relatively few things, but you must do them the SAT way.

You: What is the math SAT anyway?

Me: It is a reading test, a speed test, a trick test, but not really a math test. And you really don't need to know a zillion facts.

You: So how is this book written?

Me: The book is divided into bite-size portions, first with instructions, then with practice problems. Then there are practice SATs at the end to see how you're doing.

You: I understand calculators can be used on the SAT! Yay!!!

Me: Stop cheering. If you look at the top of each math section, you see stuff you need for the test. If you need to look at these formulas, you will never, NEVER do well on this test.

The same is almost true for calculators. A calculator may (just may) give you one or two answers, but it will slow you down so much you probably won't finish all the questions in a section. Let me show you an example:

$$\frac{2}{3} \times \frac{3}{4} \times \frac{4}{5} \times \frac{5}{6} \times \frac{6}{7} \times \frac{7}{8} \times \frac{8}{9} \times \frac{9}{10} \times \frac{10}{11} \times \frac{11}{12} \times \frac{12}{13} \times \frac{13}{14}$$

If you use a calculator, it will take nearly forever. Buut if you remember your arithmetic and cancel. . . .

$$\frac{2}{\cancel{3}} \times \frac{\cancel{3}}{\cancel{4}} \times \frac{\cancel{4}}{\cancel{5}} \times \frac{\cancel{5}}{\cancel{6}} \times \frac{\cancel{6}}{\cancel{7}} \times \frac{\cancel{7}}{\cancel{8}} \times \frac{\cancel{8}}{\cancel{9}} \times \frac{\cancel{9}}{\cancel{10}} \times \frac{\cancel{10}}{\cancel{11}} \times \frac{\cancel{11}}{\cancel{12}} \times \frac{\cancel{12}}{\cancel{13}} \times \frac{\cancel{13}}{14} = \frac{2}{14}$$

You can do all the canceling in your head, and get 2/14 = 1/7 in about 8 seconds!!!!

You: That is really neat!! I'm ready!!! Let's get started!!!!

CHAPTER 1

FRACTURED FRACTIONS

The SAT loves fractions but virtually never asks you to do pure computational problems; you know, the ones you need a calculator for. You must know what fractions are and how to compare them. This is how I usually begin.

Suppose I'm 6 years old. Can you please tell me what 3/7 is? Remember, I don't know division. So I can't change a fraction to a decimal. Heck, I don't even know what a decimal is.

OK, I'm a smart 6 year old. Suppose I have a pizza pie. . . . That is right. You divide it into 7 EQUAL parts, and I get 3 pieces.

Words and Symbols You Need

> *Sum:* **Answer in addition.**
> *Difference:* **Answer in subtraction.**
> *Product:* **Answer in multiplication.**
> *Quotient:* **Answer in division.**

a > b, read *a* is greater than b, is the same as b < a, read *b* is less than a. Also, negatives "reverse." 2 < 3 buuut −2 > −3.

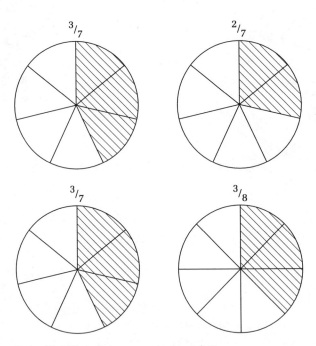

But which is bigger, 2/7 or 3/7?

3/7 is bigger: 3/7 > 2/7 because 3 equal pieces are more than 2.

But which is bigger 3/7 or 3/8?

Again, it is 3/7 because 3 larger pieces are more than 3 smaller pieces.

For positive fractions, if the BOTTOMS are the same, the bigger the top, the BIGGER the fraction. If the TOPS are the same, the bigger the bottom, the SMALLER the fraction.

For negative fractions, the opposite is true, although the SAT rarely compares negative fractions.

Trick for **adding fractions:**

$$\frac{a}{b} - \frac{c}{d} = \frac{ad - bc}{bd}$$

Reason:

$$\frac{a}{b} - \frac{c}{d} = \frac{ad}{bd} - \frac{bc}{bd} = \frac{ad - bc}{bd}$$

$$\frac{3}{10} - \frac{2}{11} = \frac{33 - 20}{110}$$

To **mulllltiply fractions,** always cancel before you multiply the tops and the bottoms:

$$a \times \frac{b}{c} = \frac{ab}{c}$$

Reason:

$$a \times \frac{b}{c} = \frac{a}{1} \times \frac{b}{c} = \frac{ab}{c}$$

To **double a fraction,** double the top or take ½ of the bottom:

Twice 5/7 is 10/7 and twice 7/6 is 7/3 (or 14/6, which = 7/3).

Here is one the SAT loves and is frequently overlooked, comparing a positive fraction to 1. This sounds simple but is not quite that.

If you add the same positive number to a fraction less than 1, it gets bigger:

$$\frac{3}{4} < \frac{3+6}{4+6} = \frac{9}{10}.$$

If you add the same positive number to a fraction bigger than 1, it gets smaller:

$$\frac{3}{2} > \frac{3+1}{2+1} = \frac{4}{3}.$$

LET'S TRY SOME PROBLEMS

The answers are explained on the next page. Do not get discouraged. Most people get most of the answers wrong the first time or two through. Reread each problem until you understand the skill. Remember, only the day of the SAT counts. By then you'll do fine.

EXAMPLE 1

$$\frac{a}{b} - \frac{a}{d}$$

EXAMPLE 2—

Quantitative analysis. This is a section that you must do quickly and accurately.

Put A if A is bigger

B if B is bigger

C if A = B

D if you can't tell

now Example 2

A. $\dfrac{1/2}{1/4}$ B. $\dfrac{1/4}{1/2}$

NOTICE

Any time there are just two choices, A and B, in this book, it is a quantitative analysis question.

EXAMPLE 3—

A. $(1/.06)^2$ B. $1/6$

EXAMPLE 4—

A. $\dfrac{a}{b}$ B. $\dfrac{a+1}{b+1}$

EXAMPLE 5—

$\dfrac{1}{1+\dfrac{1}{b}} = p.$ Which expression is 2p?

A. $\dfrac{2}{2+\dfrac{2}{b}}$ B. $\dfrac{2}{2+\dfrac{1}{b}}$ C. $\dfrac{1}{\dfrac{1}{2}+\dfrac{1}{2b}}$

D. $\dfrac{1}{1+\dfrac{2}{b}}$ E. $\dfrac{1}{2+\dfrac{1}{2b}}$

EXAMPLE 6—

If x ranges from .0002 to .02 and y goes from .2 to 20, what is the maximum value of x/y?

SOLUTIONS

EXAMPLE 1—

This should be simple:

$$\frac{ad - ab}{bd}$$

EXAMPLE 2—

You nevvvver use a calculator for this one. A is bigger because it has a bigger top and a smaller bottom than B.

EXAMPLE 3—

It is truly amazing how many get this one wrong. Again, no calculator is needed. A is bigger. A is a fraction that is bigger than 1. If you square it, it becomes even bigger. B is less than 1. How much bigger is A? It may shock you.

$$\left(\frac{1}{.06}\right)^2 = \frac{1}{.0036} = \frac{1.0000}{.0036} = \frac{10000}{36} \qquad \frac{1}{6} = \frac{6}{36}$$

A is over 1000 times bigger. Imagine!!

EXAMPLE 4—

Tough. D, you can't tell.

1. If $a = b$ (the problem doesn't say it can't)

$$\frac{a}{b} = \frac{a + 1}{b + 1} \qquad \frac{5}{5} = \frac{6}{6}$$

2. If $a < b$ $\dfrac{a}{b} < \dfrac{a + 1}{b + 1}$ $\dfrac{2}{3} < \dfrac{3}{4}$

3. If $a > b$ $\dfrac{a}{b} > \dfrac{a + 1}{b + 1}$ $\dfrac{5}{4} > \dfrac{6}{5}$

and we didn't even look at negative fractions!!!!

EXAMPLE 5—

Tuff. A is = to p because we doubled both the top and bottom. B, D, E are mixes. Answer is C because we took half of the bottom.

EXAMPLE 6—

Any positive fraction is largest with the biggest top and the smallest bottom. $.02/.2 = 1/10 = .1$, because the question was really a multiple choice one, but I wanted you to answer it without the choices since there are nonmultiple choice answers.

If you noticed, 3 of the 6 questions are comparisons. It is vital that you can do these well. In the next section there will be more general hints.

CHAPTER 2

CAN WE COMPARE? (YES!) THE POWERS THAT BE

Which is bigger a^2 or a? The answer is, we can't tell, but it is important to know why.

If $a > 1$, then $a^2 > a$ $3^2 > 3$.

If $a = 1$, then $1^2 = 1$.

Because a^2 is bigger than a in the first case and equal to a in the second, the answer must be D. But we need to do more.

If $0 < a < 1$, $a^2 < a$ since $(\frac{1}{2})^2 = \frac{1}{4} < \frac{1}{2}$.

I'll bet some of you didn't know that when you square a number, sometimes it gets smaller. This is how the SAT gets you, but not now!!!

If $a = 0$, $a^2 = a$ since $0^2 = 0$.

If $a < 0$ (negative), a^2 is bigger because squaring a number makes it positive, which is always bigger than a negative.

$(-4)^2 > -4$ since $16 > -4$.

THIS IS VERY IMPORTANT**

1. If a problem severely restricts the location of a number or numbers, the answer is more likely A or B. For example: If a > 23 (could have been 1, but the SAT is tricky), $a^2 > a$.

2. If there is no restriction on the letters and there are only letters in the problem, the most likely answers are D and C, in that order.

3. If there are only numbers in the problem, the answer is never D. It is A, B, or C. Maybe you can't tell, but someone can.

4. Mixed numbers and letters and some restrictions could be anything.

NOTE

If we compared a^3 and a^2, everything would be the same exceptttt if a < 0, $a^2 > a^3$ because a positive is > a negative. $(-4)^2 > (-4)^3$, $16 > -64$.

There are a zillion problems about this on the SAT (slight exaggeration). Let's do a few.

PROBLEMS

EXAMPLE 1—

$-43 < x < -10$.

A. $1/x^6$ B. $1/x^7$

EXAMPLE 2—

$-2 < x < -1$.

A. x   B. 1/x

EXAMPLE 3—

A. ab   B. bc

EXAMPLE 4—

A. 0.5m + 0.7p   B. 0.7(m + p)   (The SAT loves this kind.)

EXAMPLE 5—

A. average (arithmetic mean) of a, b, c   B. arithmetic mean of a^2, b^2, c^2

SOLUTIONS

EXAMPLE 1—

Lots of things to try to trick you: $-43 < x < -10$. We only need x negative reciprocals also to try to fool you. The only thing you need to know is x^6 is positive and x^7 is negative. Answer is A.

EXAMPLE 2—

Take a number between -2 and -1, say $x = -3/2$. $1/x = -2/3$

$-2/3$ is bigger. Answer is B.

EXAMPLE 3—

A realllly tricky one. If a, b, c were all positive, ab would be less than bc. But, let $a = -3$, $b = 0$, $c = 5$. $ac = bc$ aaannnd $a = -5$, $b = -2$, $c = 6$ $ac > bc$. Answer is D.

EXAMPLE 4—

Again, you can't jump to any conclusions. If you multiply out B, you get $0.7m + 0.7p$. Cancelling the 0.7p's, we are comparing 0.5m and 0.7m. 0.7m is bigger iffff m > 0. If m = 0, they are equal. If m < 0, 0.5m > 0.7m, and the answer is again D.

EXAMPLE 5—

Again the answer is D, because if you square a number, you can make it larger or smaller, or remain the same.

CHAPTER 3

ROOTS, LIKE SQUARE, MAN

The SAT also adores square roots. The SAT never asks you to calculate the square root of 123456789.234, especially with calculators, but you should know the following:

1. $\sqrt{2} = 1.4$ (approx) and $\sqrt{3} = 1.7$ (also approx).

2. $\sqrt{0} = 0$, $\sqrt{1} = 1$, $\sqrt{4} = 2$, $\sqrt{9} = 3$, $\sqrt{16} = 4$, $\sqrt{25} = 5$, $\sqrt{36} = 6$, $\sqrt{49} = 7$, $\sqrt{64} = 8$, $\sqrt{81} = 9$, $\sqrt{100} = 10$. Even with calculators, it is good to know these.

3. $\sqrt{\dfrac{4}{9}} = \dfrac{\sqrt{4}}{\sqrt{9}} = \dfrac{2}{3}$

4. Simplify $\sqrt{72}$: $\sqrt{(2)(2)(2)(3)(3)} = (3)(2)\sqrt{2} = 6\sqrt{2}$. Also . . . $10\sqrt{72} = 10(6)\sqrt{2} = 60\sqrt{2}$.

5. $2\sqrt{3} + 4\sqrt{5} + 6\sqrt{3} + 7\sqrt{5} = 8\sqrt{3} + 11\sqrt{5}$.

6. Rationalize the denominator:

$$\frac{8}{\sqrt{6}} = \frac{8\sqrt{6}}{\sqrt{6}\sqrt{6}} = \frac{8\sqrt{6}}{6} = \frac{4\sqrt{6}}{3}.$$

7. $(a\sqrt{b})(c\sqrt{d}) = ac\sqrt{bd}$.

Reason: 8. is true because if you square $\sqrt{x+y}$, you get $x + y$. If you square $\sqrt{x} + \sqrt{y}$, you get $x + y +$ a middle term.

8. x, y positive $\quad \sqrt{x} + \sqrt{y} > \sqrt{x+y}$.

And just like comparing a and a^2, we want to compare a and $\sqrt{a}$.

Just like before, the answer is D, we can't tell.

If $a > 1$, $a > \sqrt{a}$ since $4 > \sqrt{4}$.

$a = 1 \quad a = \sqrt{a}$ since $1 = \sqrt{1}$.

This is a demonstration like p12 on the top. I'll bet some of you didn't know that sometimes, when you take a square root, the number becomes bigger.

$0 < a < 1 \quad \sqrt{a} > a$!!!! $\quad \sqrt{\frac{1}{4}} = \frac{1}{2} > \frac{1}{4}$

If $a = 0$, $\sqrt{0} = 0$ and $a = \sqrt{a}$.

a can't be negative because a is imaginary, a no-no for the SAT, at least up to now.

LET'S TRY SOME PROBLEMS

EXAMPLE 1

A. $\sqrt{12}$ B. $\sqrt{5} + \sqrt{7}$

EXAMPLE 2

If $0 < M < 1$, which of the following are true?

   I. $M > M^3$

   II. $M^2 > 1/M^2$

   III. $M > 1/\sqrt{M}$

A. I only   B. II only   C. I and II   D. III only
E. I and III

EXAMPLE 3

$(\sqrt{32} + \sqrt{18})^2$

EXAMPLE 4

A. the square of 13/19   B. the square root of 13/19

SOLUTIONS

EXAMPLE 1—

The answer is, of course, B, because the

$$\sqrt{x} + \sqrt{y} > \sqrt{x+y}$$
$$\sqrt{5} + \sqrt{7} > \sqrt{5+7}$$

EXAMPLE 2—

There is usually more than one of these questions on the SAT. So, we will do this one question in great detail.

I is true, because we know cubing a number between 0 and 1 makes it smaller (squaring also does this).

II is false. M is less than 1; M^2 is less than 1; $1/M^2$ is bigger than 1.

III is false. M is less than 1; so is $\sqrt{M}$; but $\dfrac{1}{\sqrt{M}}$ is bigger than 1.

The answer is A. This problem is important and should be gone over.

EXAMPLE 3—

$$\sqrt{32} = \sqrt{(2)\,(2)\,(2)\,(2)\,(2)} = 4\sqrt{2}.$$
$$18 = \sqrt{(2)\,(3)\,(3)} = 3\sqrt{2}.$$
$$4\sqrt{2} + 3\sqrt{2} = 7\sqrt{2}.$$
$$(7\sqrt{2})^2 = (7\sqrt{2})(7\sqrt{2}) = 49\sqrt{4} = 49(2) = 98.$$

EXAMPLE 4—

B, of course, because squaring a number between 0 and 1 makes it smaller, aaaannd square rooting a number between 0 and 1 makes it bigger.

More on square roots later with the Pythagorean theorem, distance formula, 45-45-90 right triangle, and 30-60-90 right triangle.

CHAPTER 4

AVERAGES

Let's try an easy topic for a change. Most students seem to like averages. Going over past SATs, I was amazed to see how many times averages occurred. The average (arithmetic mean) is the way you are graded.

EXAMPLE 1—

Find the average (arithmetic mean) of 45, 35, 20, 32.
45 + 35 + 20 + 32 = 132. 132/4 (numbers) = 33, the mean.

However, the SAT would NEVER ask the question this way, never!!! It would be something like this: Sandy received 76 and 89 on two tests. What must be the third test score for Sandy to have an 80 average?

METHOD 1

80 average on three tests means 3 × 80 = 240 points. So far, Sandy has 76 + 89 = 165 points. Sandy needs 240 − 165 = 75 points.

METHOD 2

80 average. 76 is −4 points; 80 is +9 points. So far +5 points. Needed is 80 − 5 = 75 points, since +5 − 5 = 0.

NOTE: The SAT always says *arithmetic mean* because there are 2 other words that mean average:

  median—the middle number,
  mode—the most common

  We arrange the following nine numbers in order:

2, 3, 3, 3, 7, 12, 14, 15, 19

The median, the middle grade, is the fifth of the nine numbers, or 7.

We arrange the following ten numbers in order:

2, 3, 3, 3, 4, 7, 12, 14, 15, 19

The median, the middle grade, is the mean of the middle two numbers, the mean of the fifth and sixth (4 + 7)/2 = 5.5, the median.

In each case the mode, the most common, is 3.

Of the three words meaning average, the most accurate of a pretty large or large group is the MEDIAN.

I like method 2; most students like method 1. Choose the one you like.

LET'S TRY SOME PROBLEMS

EXAMPLE 2—

$6x + 6y = 48$. Find the average (arithmetic mean) of x and y.

EXAMPLE 3—

In a certain city, the arithmetic mean of high readings for 4 days was 63°F. If the high readings were 62°F, 56°F, and 68°F for the first 3 days, what was the temperature the fourth day?

EXAMPLE 4—

On a certain test, a class with 10 students has an average (arithmetic mean) of 70, and a class of 15 has an average of 90. What is the average of the 25 students?

EXAMPLE 5—

On a certain test, the sophs had an 82 average and the juniors a 92 average.

A. average of the total group   B. 87

SOLUTIONS

EXAMPLE 2—

The mean of x and y is (x + y)/2.

$6x + 6y = 48$

Soooo $x + y = 8$ (divide by 6)

Aaaand $\dfrac{x + y}{2} = 4$ (divide by 2)

EXAMPLE 3—

METHOD I
$63 \times 4 = 252°$. $62 + 56 + 68 = 186°$. $252 - 186 = 66°F$.

NOTICE
$186°$ total makes no sense, but who cares.

METHOD 2
Average $63°F$

$62°F$ is $-1°$.

$56°F$ is $-7°$.

$68°F$ is $+5°$.   So far, $-3°$ $(-1 - 7 + 5 = -3)$

Soo, the fourth day is $63 + 3 = 66°F$.

NOTICE

It wouldn't make a difference if it were Celsius.

EXAMPLE 4—

The average is NOT NOT NOT 80 because the groups are not equal.

$$10(70) = \ \ 700 \text{ points}$$
$$15(90) = 1350 \text{ points}$$

Totals: 25   2050 points

Average: $\dfrac{2050}{25} \times \dfrac{4}{4} = \dfrac{8200}{100} = 82$ average.

NOTICE
You never divide by 25. You multiply by 4 and divide by 100, because 25 is 100/4, and both multiplying by 4 and dividing by 100 is way easier than dividing by 25.

EXAMPLE 5—

For exactly the same reason as in Example 4, the answer is D, can't tell.

If the groups were equal size, the average would be 87.

More sophs? The average would be less than 87.

More juniors? The average would be more than 87.

CHAPTER 5

A LITTLE NUMBER THEORY: ODDS, EVENS, PRIMES

Another topic is number theory. No, no, no!! Not high-level stuff. I think this is the fun stuff. You need to know (if you don't already know) the following:

Integers: −5, −4, −3, −2, −1, 0, 1, 2, 3, 4, 5. . . .

Positive integers: 1, 2, 3, 4, 5. . . .

Evens: −6, −4, −2, 0, 2, 4, 6. . . . (remember 0 and negs)

Odds: −7, −5, −3, −1, 1, 3, 5, 7. . . .

Multiples of 3: −9, −6, −3, 0, 3, 6, 9. . . . etc.

YOU MUST KNOW THESE *WELL*. If you do not know these, put in integers and convince yourself that:

even + even = even   even + odd = odd
odd + odd = even

even × even = even   even × odd = even
odd × odd = odd

If n is an integer. . . .

n + 1 is the next consecutive integer; n + 2 is the one after that, and so on.

WARNINGS***WARNINGS

If the problem does NOT say integers, then it can be any real number: fraction, irrational, any real number.

Specifically: a > 0 (a positive) can mean any decimal number, rational or irrational.

The sum of three consecutive integers is $n + n + 1 + n + 2 = 3n + 3$.

If n is an even integer, . . . $n + 2$ is the next, $n + 4$ is the next ($n - 2$ is the even number before n).

If n is an odd integer, . . . $n + 2$ is the next!!!! (5 is odd, $5 + 2 = 7$ is the next consecutive odd, just like the evens), the next consecutive odd is $n + 4$.

Squaring

If n is odd, n^2 is odd; aaannnd if n^2 is odd and n is an integer, n is odd. $3^2 = 9$ and backward.

If n is even, n^2 is even; aannd if n^2 is even and n is an integer, n is even. $6^2 = 36$ and backward.

Factors (also called *divisors*) are integers that "go into" a number with no remainder. We'll only look at positive ones.

Factors of 30: 1, 2, 3, 5, 6, 10, 15, 30

NOTICE

2 is the only even prime.

Primes are positive integers with exactly two distinct factors: itself and 1.

Prime factors of 30: 2, 3, 5

It might be nice to know the first eight: 2, 3, 5, 7, 11, 13, 17, 19.

NOTICE

1 is <u>not</u> a prime. It has only one distinct factor, namely 1!!!!

I love problems on this page. So does the SAT. Let's do a bunch!!!!

PROBLEMS

EXAMPLE 1—

n is an integer. Which is never even?

A. 2n   B. 2n + 1   C. 3n + 2   D. 2(n − 1)
E. $2(n + 1)^3$

EXAMPLE 2—

The sum of two consecutive positive integers is never divisible by

A. 2   B. 3   C. 5   D. 9   E. 799

EXAMPLE 3—

m is odd and n is even. Which could be even?

A. m + n   B. m − n   C. m/2 + n   D. m + n/2
E. m/2 + n/2

EXAMPLE 4—

If the arithmetic mean (average) of 10 consecutive integers is 15½, arranged in increasing order, what is the mean of the first five?

EXAMPLE 5—

$$\frac{(M − 2)(M − 4)(M − 6)(M − 20) − 1}{2}$$ is an integer if M = ?

EXAMPLE 6—

If q is odd, which is even?

A. q/2   B. q + 2   C. 2q + 1   D. q^3 E. q(q + 1)

EXAMPLE 7—

Which letter shows that not all odds are primes?

A. 3   B. 5   C. 7   D. 13   E. 21

EXAMPLE 8

T is the set of multiples of 3. T = {−9, −6, −3, 0, 3, 6, 9. . . . } If c and d are in set T, which of the following is NOT in T?

A. cd   B. c + d   C. c − d   D. −c − d   E. c/d
F. $c^2 - d$

EXAMPLE 9

If q is a whole number and q is a prime, and if 20q is divisible by 6, then q could be

A. 2   B. 3   C. 4   D. 5   E. 6

EXAMPLE 10

If two odd integers are primes, which of the following is true?

    A. Their product is an odd integer.

    B. Their sum is prime.

    C. Their sum is an odd integer.

    D. Their product is prime.

    E. The sum of their squares is prime.

    F. Their product added to 1 is prime.

    G. The sum of their cubes is prime.

EXAMPLE 11

y is an even integer.

A. distinct prime factors of y
B. distinct prime factors of 2y

EXAMPLE 12

The sum of the first m positive integers is y. In terms of m and y, which of the following is the sum of the next m positive integers?

A. my   B. m + y   C. m² + y   D. m + y²
E. 2m + y   F. m + 2y

NOW FOR THE ANSWERS. . . .

EXAMPLE 1—

C can be either odd or even (try n = 1 and n = 2). E, D,
A are always even because they are multiples of 2. If
2n is always even, 2n + 1 is always odd.   B.

EXAMPLE 2—

The sum of two consecutive integers is always odd (odd +
even is odd) and is never divisible by 2. Answer is A. (A
little tricky, but no hard choices to throw you off.)

EXAMPLE 3—

m/2 is a fraction, C and E are no good! A and B are always
odd. Answer is D. (If m = 7 and n = 22, for example.)

EXAMPLE 4—

15½ is the middle of 10 integers (consecutive type).
Therefore, there are 5 less and 5 larger. 5 less must be
15, 14, 13, 12, 11. The average of an odd number is the
middle: 13.

EXAMPLE 5—

This is very tough. If M is an even integer, (M − 2) ×
(M − 4)(M − 6)(M − 20), this product will always be
even. Even −1 is odd, divided by 2 is a fraction,
always. If M is odd, the product of (M − 2)(M − 4)(M −
6)(M − 20) is odd; any odd integer minus 1 is an even
integer; any even integer divided by 2 is always an
integer. The answer is any and all odd integers.

EXAMPLE 6—

q/2 is a fraction; q + 2 = next consecutive odd; 2q + 1 is
always odd; q³ like q² is odd. Answer is E. If q is odd,

then q + 1 is even. The product of an odd and even is always even.

EXAMPLE 7—

Easy one, E. 7(3) = 21.

EXAMPLE 8—

I know, I know, the SAT has only five choices, but this is a learning process. Take c = 6 and d = 9. You will find only E is not a multiple of 3. This is true for even numbers or multiples of anything.

EXAMPLE 9—

A, B, D are primes. If q = B or E, 20q is divisible by 6. Only B fits both categories.

EXAMPLE 10—

Again, the SAT doesn't have seven choices. If you take 3 and 5, only A is OK. Actually, try all of them to see what happens. Remember, this is only practice.

EXAMPLE 11—

A toughie. If y is even and has prime factors 2, p, q, . . . then, if you multiply by 2, the factors increase but the primes don't. Answer is C.

12 has factors 1, 2, 3, 4, 6, 12; prime factors 2, 3.

24 has factors 1, 2, 3, 4, 6, 8, 12, 24; prime factors 2, 3.

However, if y had been odd, 2y, choice B, would be larger because 2y would have one extra prime factor, 2.

EXAMPLE 12—

This is truly a hard one, one of the few in this category. Let us say m = 6. Then y = 1 + 2 + 3 + 4 + 5 + 6. The next six would be 7 + 8 + 9 + 10 + 11 + 12. Looking at these differently, $7 + 8 + 9 + 10 + 11 + 12 = (1 + 6) + (2 + 6) + (3 + 6) + 4 + 6) + (5 + 6) + (6 + 6) = 1 + 2 + 3 + 4 + 5 + 6 + 6(6) = y + 6^2$. In general, $y + m^2$.   C.

NOTE 1
You should never worry about problems you can't do or that are strange. If you get all the ones correct that you know how to do, you'll do very fine!!!

NOTE 2
Never stay on one problem too long. Go on to the next and guess if you run out of time.

NOTE 3
Once you have answered any question, forget it and go on to the next question. NEVER, NEVER, NEVER change a question unless you are 100% (not 99%) sure you are correct. The last time I took a test, I broke my own rule; and I changed a right answer to a wrong one.

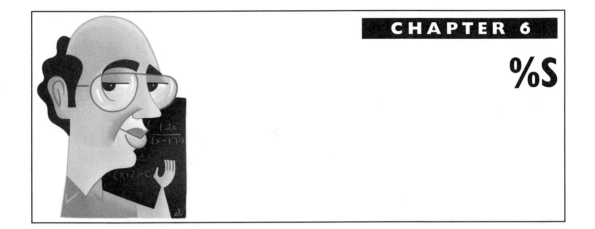

CHAPTER 6

%S

Let's try to finish up the arithmetic. The biggest problem is usually %. Let's cut it down to size.

You should know the following equivalents, even with a calculator.

$1\% = .01 = 1/100$ $10\% = .10 = 1/10$ $20\% = .20 = 1/5$
$25\% = .25 = 1/4$ $30\% = .30 = 3/10$ $40\% = .40 = 2/5$
$50\% = .50 = 1/2$ $60\% = .60 = 3/5$ $70\% = .70 = 7/10$
$75\% = .75 = 3/4$ $80\% = .80 = 4/5$ $90\% = .90 = 9/10$
$100\% = 1.0 = 1$ $200\% = 2.0 = 2$

You should also know the following:

$1/6 = .1\overline{6} = 16\frac{2}{3}\%$ $1/3 = .\overline{3} = 33\frac{1}{3}\%$
$2/3 = .\overline{6} = 66\frac{2}{3}\%$ $5/6 = .8\overline{3} = 83\frac{1}{3}\%$
$1/8 = .125 = 12\frac{1}{2}\%$ $3/8 = .375 = 37\frac{1}{2}\%$
$5/8 = .625 = 62\frac{1}{2}\%$ $7/8 = .875 = 87\frac{1}{2}\%$

Even if you have been terrrrible with %s, I'll bet you'll get it now. It is the pyramid method. The goal is to get the pyramid in your head. It shouldn't be too hard.

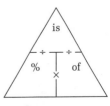

RECALL: $.5\overline{67} = .5676767....$

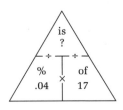

EXAMPLE 1—

4% of 17 is what?

.04 × 17 = .68

> 4% = .04 goes into % box.
>
> 17 goes into the "of" box.
>
> The chart says multiply.

That's it.

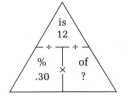

EXAMPLE 2—

12 is 30% of what?

$$\frac{12}{.30} = \frac{120}{3} = 40$$

> 12 in the "is" box.
>
> 30% = .30 in % box.

Chart says divide.

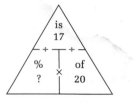

EXAMPLE 3—

17 is what % of 20?

$$\frac{17}{20} = 20\overline{)17.00} \quad .85 = 85\%$$

> 17 in the "is" box.
>
> 20 in the "of" box.

Pyramid says divide.

REMEMBER

THE PYRAMID SHOULD BE IN YOUR HEAD!

EXAMPLE 4—

A television costs $400. If the sales tax is 8%, how much do you pay? Tax is 8% of $400. Soooo

.08 × $400 = $32 $400 + $32 = $432

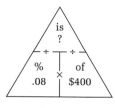

Simple interest problems are the same.

Principle = $400.
Interest rate = 8%.
Interest = $400 × .08 = $32.
Total you have is $400 + $32 = $432

EXAMPLE 5—

A $50 dress is discounted 15%. How much do you pay? Discount is on the original price. Sooo

$50 × .15 = $7.50

You pay $50.00 − $7.50 = $42.50

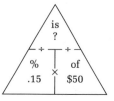

LET'S TRY SOME

Let's do some purely arithmetic % problems.

EXAMPLE 1—

If 15 kg of pure H_2O are added to 10 kg of pure alcohol, what % by weight of the resulting solution is alcohol?

EXAMPLE 2—

The weight of a 2000 ton truck is increased by 1%. What is the weight in tons of the increased load?

EXAMPLE 3—

In a certain widget factory, .08% are defective. On the average, 4 will be defective. How many widgets are produced?

EXAMPLE 4—

In a certain country, the ratio of people over 40 to people under 40 is 3 to 2. What % of the population is under 40?

EXAMPLE 5—

What is 50% of 50% of 50% of 1?

SOLUTIONS

EXAMPLE 1—

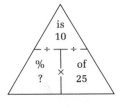

Alcohol is 10 kg out of a 25-kg total.

$$\frac{10}{25} = \frac{2}{5} = 40\%$$

EXAMPLE 2—

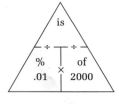

$2000 \times .01 = 20$. $2000 + 20 = 2020$. The "tons" is there to throw you off.

EXAMPLE 3—

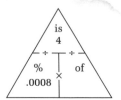

You should read this problem as

.08% of what (total) is 4?

.08% is not a decimal. It is a %.

.08% = .0008 (2 decimal places to the left)

$$\frac{4}{.0008} = \frac{40,000}{8} = 5000 \text{ widgets}$$

EXAMPLE 4—

Hopefully, you see that this is exactly Example 1 (2 out of 5 are under 40).

EXAMPLE 5—

This is easier done in fraction form. Repeated "ofs" and % box means repeated multiplication.

½ × ½ × ½ × 1 = 1/8

% will show up later when we do charts, reading problems, ratio, and algebra. These should show you that you can do % problems!!!!!

Speaking of algebra, let's do some.

CHAPTER 7

SUBSTITUTE SUBSTITUTIONS

For those of you who think you have to be algebraic whizzes for the SAT and know a ton of facts, you are wrong, wrong, wrong!!! It is truly amazing how little algebra you need to do well. But you need to know how the SAT asks it. We will start with substitution (examples only) and do the rest in small pieces.

NOTE: You must know the difference between

$$-3^2, \ (-3)^2, \ -(-3)^2$$
$$-3^2 = -3(3) = -9$$
(one minus sign = −).
$$(-3)^2 = (-3)(-3) = +9$$
(two minus signs = +).
$$-(-3)^2 = -(-3)(-3) = -9$$
(three minus signs = −.)

PROBLEMS

Yes, you can do these!!

EXAMPLE 1—

$x = -3, \ y = 0 \qquad x^2y + y/x =$

EXAMPLE 2—

$(x + 4)^2 = (x - 4)^2 \qquad x =$

A. 0   B. 2   C. 4   D. 8   E. 12

EXAMPLE 3—

$x, \ y$ are positive integers and $x > y$ and $x^2 + y^2 = 13$.
$x - y =$
A. 1   B. 2   C. 3   D. 5   E. 7

EXAMPLE 4—

$5x = -3$ $(5x - 2)^2 =$

EXAMPLE 5—

$2x < 24 < 3x$. x an integer.

A. x   B. 8

EXAMPLE 6—

Of the following values for M, $(-1/4)^M$ will have the greatest value if M =

A. 2   B. 3   C. 4   D. 5   E. 10

EXAMPLE 7—

If a = 2 and b = 4, which of the following is NOT equal to the others?

A. $(a - b)^2$ B. 2a   C. $(b - a)^2$ D. $6a - 2b$ E. $8a - 6b$

EXAMPLE 8—

If x = 1 and y = -2, $2x - 3y =$

A. -7   B. -4   C. -1   D. 5   E. 8

EXAMPLE 9—

If M is the least positive integer for which 3M is both an even integer and equal to the square of an integer, then M =

A. 1/3   B. 3   C. 6   D. 12   E. 48

EXAMPLE 10—

$4 < xy < 16$

A. xy   B. x + y

EXAMPLE 11—

$y = 3x$ $x \neq 0$

A. $3x - y$ B. $3xy$

EXAMPLE 12—

8/M is an odd integer. Which of the following could be M?

A. 8/3   B. 5/8   C. 4/3   D. 1/3   E. 1/3

EXAMPLE 13—

If m/2 is even and m/4 is odd, m could be

A. 32   B. 24   C. 20   D. 16   E. 10

EXAMPLE 14—

If x + y = 2, then x + y − 3 =

A. 5   B. 3   C. 1   D. −1   E. −3

SOLUTIONS

EXAMPLE 1—

Pure substitution.

$(-3)^2(0) = 0$ $0/(-3) = 0$ $0 + 0 = 0$

A. Careful with zeroes.

EXAMPLE 2—

Trial and error. 0 works since $(0 - 4)^2 = (0 + 4)^2$.   A.

EXAMPLE 3—

Again, trial and error. x = 3 and y = 2 satisfies both conditions. x − y = 1.   A.

EXAMPLE 4—

Don't solve for x!!!! 5x = −3. Soooo $(5x - 2)^2 = (-3 - 2)^2 = (-5)^2 = 25$.

EXAMPLE 5—

x can't be 8 because 24 is not < 24. x = 9, 10, 11 work. All are bigger than x. Answer is A!!

EXAMPLE 6—

M can't be B or D because a negative is always smaller than a positive. Also, numbers between 0 and 1, when raised to a power, get smaller. Therefore, the smallest even value for M is correct. A is the answer.

NOTICE

On this set, A occurs very often as the answer. Years ago, when test makers didn't know as much, C or D were the most common answers. Now, anything is possible. Let's go on.

EXAMPLE 7—

E is the only one where the answer is not 4. (It is −8.) Try to do this kind very quickly.

EXAMPLE 8—

Pure substitution. Be careful. $2(1) - 3(-2) = 2 + 6 = 8$.   E.

EXAMPLE 9—

A, D, E fit one condition, but A is not an integer and D is smaller than E. Answer, D.

EXAMPLE 10—

Trial and error. If $x = 1$ and $y = 5$, $xy < x + y$. If $x = 2$ and $y = 4$, $xy > x + y$.   Answer D, can't tell.

EXAMPLE 11—

If $y = 3x$, $3x - y = 0$. If $y = 3x$, $3xy = 3x(3x) = 9x^2 > 0$ since $x \neq 0$.   Answer, B.

EXAMPLE 12—

B means $8/M$ is not even an integer. C, D, E result in an even. Answer, A.

EXAMPLE 13—

m/2 is odd and m/4 even??? Only C satisfies both conditions.

EXAMPLE 14—

Do not solve for x or y. Substitute x + y = 2. x + y − 3 = 2 − 3 = −1.   D.

This batch is not too bad. Let's try something else.

I believe with the change in exam and the use of calculators, this type of problem will increase a lot!!!! So, we will spend some time here. Most of these problems are short. The difficulty occurs because some of these questions are not done in school.

1. You must know how to compute, preferably without a calculator.

   a. $2^3 = 2(2)(2) = 8$ b. $3^{-4} = 1/3^4 = 1/81$

2. a. $x^m x^n = x^{m+n}$ b. $x^6 x^3 x = x^{10}$
   c. $(3x^4 y^5)(-2x^{11} y^2) = -6x^{15} y^7$
   d. $x^{cats} x^{dogs} = x^{cats + dogs}$ e. $2^9 2^5 = 2^{14}$
   f. $x^8 x = x^{8+1}$

3. a. $x^m / x^n = x^{m-n}$ b. $x^{food} / x^{calories} = x^{food - calories}$
   c. $\dfrac{4x^4 y^5 z^6}{6x^9 y^5 z^2} = \dfrac{2z^4}{3x^5}$ d. $\dfrac{x^a x^c}{x^e} = x^{a+c-e}$

4. a. $(x^m)^n = x^{mn}$ b. $(x^5)^3 = x^{15}$
   c. $(5xy^6)^2 = 25x^2 y^{12}$ d. $(x^{4\ bananas})^{3\ bananas} = x^{12\ bananas^2}$

Basics are fairly straightforward. However, some of these can be a little tricky. Let's try some.

NOTES:

1b. Negative exponents mean reciprocal: nothing to do with negative numbers.

2. When you multiply, if the base is the same, add the exponents.

2b, f. $x = x^1$.

2e. If the base is the same, add exponents, buuut base stays the same.

3. When you divide, you subtract exponents.

3c. $y^5/y^5 = 1$, and if the larger exponent is on the bottom, the answer is on the bottom. $x^4/x^9 = 1/x^5$. The 1 is not needed unless the top is all cancelled out.

3d. Combine 2 and 3.

4a. Power to a power? Multiply exponents.

4b. $(x^5)^3 = x^5x^5x^5 = x^{15}$.

PROBLEMS

EXAMPLE 1

A. $10^{60} - 10$ B. $10^{59} + 10$

EXAMPLE 2

$\dfrac{a^{x+y}}{a^x} =$ A. a^y B. $1/a^y$ C. $-a^y$ D. a^{1+y}

E. $1 + a^y$

EXAMPLE 3

$\left(\dfrac{x^6y^5}{x^3y^2}\right)^2 =$ A. x^9y^9 B. x^9y^5 C. x^6y^6 D. x^5y^5

E. x^4y^6

EXAMPLE 4

$\dfrac{x^{2a+3}x^{4a+7}}{x^c}$

EXAMPLE 5

$3 = b^y$. Then $3b =$ A. b^{y+1} B. b^{y+2} C. b^{y+3}
D. b^{2y} E. b^{3y}

EXAMPLE 6

$\dfrac{1}{10^{29}} - \dfrac{1}{10^{30}}$

EXAMPLE 7

$3^n + 3^n + 3^n$

SOLUTIONS

EXAMPLE 1

10^{60} is muuuch bigger than 10^{59} (10 times a huge number). Adding or subtracting 10 means almost nothing. A.

EXAMPLE 2—

This is hard only because some of you have never seen this in school. Now, it should be easy. . . . $a^{x+y-x} = a^y$.   A.

EXAMPLE 3—

Simplify inside parentheses first. $(x^3y^3)^2 = x^6y^6$.   C.

EXAMPLE 4—

Base is the same; when you multiply, you add exponents, and when you divide, you subtract exponents.

$$x^{2a+3+4a+7-c} = x^{6a+10-c}$$

EXAMPLE 5—

This is a problem that looks like it should be easy but is not.

$3 = b^y$. Multiply both sides by b!!!! $3b = b^y b = b^{y+1}$.   A.

EXAMPLE 6—

We have to go back to fifth grade and take a look at a problem.

$$\frac{7}{16} - \frac{3}{8}$$

How is this problem similar? You'llll seeee.

LCD 16 $\dfrac{7}{16} - \dfrac{6}{16} = \dfrac{1}{16}$

Let's look at it a different way. LCD $= 16 = 2^4$. $8 = 2^3$
$2^4 = 2(2^3)$.

Multiply second fraction, top and bottom, by 2.

$$\frac{7}{2^4} - \frac{3}{2^3} = \frac{7}{2^4} - \frac{3(2)}{2^3(2)} = \frac{7-6}{2^4} = \frac{1}{2^4}$$

Sooooo $\dfrac{1}{10^{29}} - \dfrac{1}{10^{30}} = \dfrac{1(10)}{10^{29}(10)} - \dfrac{1}{10^{30}} = \dfrac{9}{10^{30}}$

EXAMPLE 7—

A truly difficult problem. It is difficult because it is addition, not multiplication.

$$3^n + 3^n + 3^n = 1(3^n) + 1(3^n) + 1(3^n) = 3(3^n) = 3^1 3^n = 3^{n+1}$$

Fortunately, few are like this one. Let's do some more algebra.

CHAPTER 9

DISTRIBUTIVE LAW, FACTORING, REDUCING, ODDS AND ENDS

1. Let's do odds and ends first—**combining like terms.**

   a. $3a + 5b + 7a + 9b$ Answer: $10a + 14b$.
   b. $4x^2 - 7x - 9 + -7x^2 + 7x - 3$ Answer: $-3x^2 - 12$.

   In combining like terms, add or subtract; leave the exponents alone. Unlike terms, different letters, or the same letter(s) with different exponents, cannot be combined.

2. Next, the basic **distributive law.**

   a. $5(2x - 3) = 10x - 15$ b. $-2(5x - 7) = -10x + 14$
   (carrrreful of 2nd sign)   c. $4(5x - 3) - 7(4x - 1)$
   $= 20x - 12 - 28x + 7 = -8x - 5$

3. You need to be able to **multiply a binomial** kwikkkly by the F.O.I.L. method in your head!!!
   (F.O.I.L. = F = First, O = Outer, I = Inner, L = Last)

   a. $(x + 5)(x - 3)$ First $x(x) = x^2$; Outer $x(-3) = -3x$; Inner $5(x) = 5x$; Last $(5)(-3) = x^2 + 2x$ (add inner and outer) -15

   b. $(x + 10)^2 = (x + 10)(x + 10) = x^2 + 20x + 100$

You must know these verrrry well. You'll see soon:

$. $(x + y)^2 = x^2 + 2xy + y^2$ $$. $(x - y)^2 = x^2 - 2xy + y^2$ $$$. $(x + y)(x - y) = x^2 - y^2$

You must know 3 kinds of factoring:

Notes:

4a. Largest number that multiplies 12 and 16 is 4; x is a factor of x and x^2.

4b. x^2 is a factor of x^2, x^3, x^4. If a whole term ($3x^2$) is factored out, a 1 goes in the parentheses.

4. Take out the **largest** common **factor.**

   a. $12x^2 - 16x = 4x(3x - 4)$

   b. $9x^4 - 15x^3 + 3x^2 = 3x^2(3x^2 - 5x + 1)$

   c. $16a^2 \text{ bug} - 24a \text{ bug} = 8a \text{ bug}(2a - 3)$

5. Difference of 2 squares,

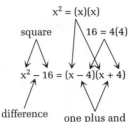

6. **Trinomials.** Verry important. We'll spend some time here because some of the newer and older books do not teach this properly. My way is the right way. Seriously, when I took algebra 400 years ago, all the teachers taught it this way. It still is the best way.

   $x^2 - 7x + 6.$ It factors into 2 binomials. Make 2 sets of ( ).

   Last sign + (+6), sign in each ( ) will be the same.

   Only if the last sign is +, look at the first sign.

   The first sign is minus (−7) means both signs are minus (first plus, both plus).

Look at first term. $x^2 = x(x)$. Temporarily ignore the middle term.

$6 = 1(6)$ or $(3)(2)$. Factors must add to 7.

The correct answer is . . . $(x - 1)(x - 6)$, where the 1 and 6 could be switched.

$a^2 - 2a - 24$.   Last sign minus (-24), means one sign in ( ) is + and one is −. $a^2 = a(a)$. $24 = 1(2) = 2(12) = 3(8) = 4(6)$. It must add to −2. 4 and −6.

$(a - 6)(a + 4)$ is the answer.

As most of you know, the first term doesn't have to be x^2. It could be $2x^2$ or $3x^2$, . . . but up to this point, if it happens, there is always a common factor, followed by a second factoring.

7. **Double factorings.**

   a. $3x^2 + 9x + 6 = 3(x^2 + 3x + 2) = 3(x + 1)(x + 2)$

   b. $2x^2 - 50 = 2(x^2 - 25) = 2(x + 5)(x - 5)$

   c. $300 - 3a^2 = 3(100 - a^2) = 3(10 + a)(10 - a)$

   If you need more help, you should get my *Algebra for the Clueless* or *Calc for the Clueless, Precalc with Trig for the Clueless* or the rest of the series.

8. **Reducing fractions.**

   a. $\dfrac{2x + 8}{2}$ Is it $x + 8$? Wrong! Is it $2x + 4$?

   Wrong! $\dfrac{2x + 8}{2} = \dfrac{2x}{2} + \dfrac{8}{2} = x + 4$, because of the order of operations.

   b. $\dfrac{x^2 - 4}{x^2 - 4x + 4} = \dfrac{(x + 2)(x - 2)}{(x - 2)(x - 2)} = \dfrac{x + 2}{x - 2}$

NOTES:

8a is the method with I term on the bottom.

8b is the method with 2 or more terms on the bottom.

LET'S TRY SOME OF THESE

I don't know about you, but I can't wait!!!!!

EXAMPLE 1—

A. $4(3x - 3)$ B. $3(4x - 4)$

EXAMPLE 2—

If $x = b - 3$ and $y = b + 6$, $x - y =$

A. -9 B. 3 C. -3 D. 9 E. $2b + 3$

EXAMPLE 3—

If $m^2 - n^2 = 18$, then $3(m + n)(m - n) =$

A. 6 B. 15 C. 24 D. 27 E. 54

EXAMPLE 4—

If $y \neq 2$

A. $\dfrac{y^2 + y - 6}{y - 2}$ B. $y + 3$

EXAMPLE 5—

If $(g + 1/g)^2 = 40$, then $g^2 + 1/g^2 =$

A. 38 B. 39 C. 40 D. 42 E. 1599

EXAMPLE 6—

A. $c^2 - d^2 + 2d(c + d)$ B. $(c + d)^2$

EXAMPLE 7—

If $a > 0$ and $a^2 - 1 = 41 \times 43$, $a =$

A. 2 B. 40 C. 41 D. 42 E. 43

EXAMPLE 8—

A. $\dfrac{12a + 4}{4}$ B. $3a$

EXAMPLE 9—

A. $\dfrac{4a^2 - 4}{a - 1}$ B. $3a + 3$

EXAMPLE 10—

What is the result if $5 + 2a$ is subtracted from the sum of $5 - a$ and $4a^2 + 5a + 6$?

EXAMPLE 11—

$ab = -2$ and $a^2 + b^2 = 16$

A. $(a + b)^2$ B. 13

EXAMPLE 12—

$a + b = m$ annnd $a - b = 1/m$, thennnn $a^2 - b^2 =$

A. $1/m$ B. m C. m^2 D. 2 E. 1

LET'S LOOK AT THE ANSWERS!!

EXAMPLE 1—

If you multiply each out, they are both $12x - 12$. C. No tricks.

EXAMPLE 2—

$x - y = (b - 3) - (b + 6) = b - 3 - b - 6 = -3 - 6 = -9.$ A.

EXAMPLE 3—

$3(m^2 - n^2) = 3(m + n)(m - n) = 3(18) = 54.$ E.

EXAMPLE 4—

Straight factoring (you must know how to do this).

$\dfrac{(y + 3)(y - 2)}{(y - 2)} = (y + 3).$

Equal, C!!

EXAMPLE 5—

This is a realllly tough one.

$(g + 1/g)^2 = (g + 1/g)(g + 1/g) = g^2 + 2g(1/g) + 1/g^2$

$= g^2 + 2 + 1/g^2 = 40$

$\quad\quad - 2 \quad\quad\quad = -2$

Soooo $g^2 + 1/g^2 = 38.$ A.

EXAMPLE 6—

A is $c^2 - d^2 + 2d(c + d) = c^2 - d^2 + 2cd + 2d^2 = c^2 + 2cd + d^2$ which you know is the same as B. Answer is C.

EXAMPLE 7—

You know $a^2 - 1 = (a + 1)(a - 1)$. You need to see $41 \times 43 = (42 - 1)(42 + 1)$.   Answer is 42, D.

EXAMPLE 8—

$\dfrac{12a + 4}{4} = \dfrac{12a}{4} + \dfrac{4}{4} = 3a + 1 > 3a.$ A.

EXAMPLE 9—

$4a^2 - 4 = 4(a^2 - 1) = 4(a + 1)(a - 1)$

Sooooooo

$\dfrac{4a^2 - 4}{a - 1} = \dfrac{4(a - 1)(a + 1)}{a - 1} = 4(a + 1) = 4a + 4$

Comparing $4a + 4$ with $3a + 3$ . . . if $a = 0$, $4a + 4$ is bigger; if $a = -10$, $-36 < -33$. Answer is D. Hopefully, this will not take you too long to do.

EXAMPLE 10—

This would be a problem with choices, but it is really mostly a reading problem.

It says: Take $4a^2 + 5a + 6$, add $5 - a$, and then subtract $5 + 2a$ (*subtracted from* reverses the order because subtract 2 from 6 means $6 - 2$).

$4a^2 + 5a + 6 + 5 - a - 5 - 2a = 4a^2 + 2a + 6.$

EXAMPLE 11—

Not an easy one.

$(a + b)^2 = a^2 + 2ab + b^2$.

buuut $a^2 + b^2 = 16$ and $ab = -2$
so $2ab = 2(-2) = -4$.

$a^2 + 2ab + b^2 = a^2 + b^2 + 2ab = 16 - 4 = 12$.

B is the answer.

EXAMPLE 12—

$$a^2 - b^2 = (a + b)(a - b) = m(1/m) = \frac{m}{1} \times \frac{1}{m} = 1 \qquad \text{E.}$$

Most of the algebraic problems are not too bad if you know your basic algebra.

Let's do some more. The next section is the part that most students like a lot, basic equations.

CHAPTER 10

EQUATIONS: ONE UNKNOWN

Let's do the basic types first. More later. The SAT seems to ask a lot of these. Most are not too hard.

Linear equations, first degree equations, can get verrrry long. On the SAT they are almost all short or very short or very very short, sometimes a little tricky, sometimes truly easy. Here are a few samples.

REMEMBER: You want to do as many steps as possible in your head!!!!

EXAMPLE 1

$3x + 19 = 4$

$ -19 = -19$

$3x \phantom{{}+19} = -15 \qquad x = -15/3 = -5$

EXAMPLE 2

$\dfrac{x}{3} + \dfrac{x}{2} = 4.$ Multiply by LCD 6.

$2x + 3x = 24. \qquad 5x = 24. \qquad x = 24/5.$

EXAMPLE 3—

$\frac{4}{x} - \frac{2}{x} = 12.$　　LCD = x.　　$4 - 2 = 12x.$

$12x = 2.$　　$x = 2/12 = 1/6.$

EXAMPLE 4—

$x + 1 + x + 3 + x + 5 = 27.$　　Combine like terms.

$3x + 9 = 27.$　　$3x = 18.$　　$x = 6.$

EXAMPLE 5—

If there are only 2 fractions . . .

$\frac{7}{9} = \frac{x}{5}$　　cross multiply.

$9x = 35$　　$x = 35/9.$　　Better . . . in your head

$x = 5(7)/9.$

Here's something more you might like to know if you have two fractions . . . (not in Example 5).

If $\frac{a}{b} = \frac{c}{d}$　　then　　$\frac{b}{a} = \frac{d}{c}$　　(flip)

$\frac{a}{c} = \frac{b}{d}$　　aaaannnd　　$\frac{d}{b} = \frac{c}{a}$

In numbers, since $3/6 = 4/2$, . . . $6/3 = 4/2$ annd $3/2 = 6/4$ annnd $4/6 = 2/3$. Thththat's allll!!!

Seriously, there isn't too much more. However the SAT sometimes is tricky. Let's try some.

PROBLEMS

EXAMPLE 1—

If $a - 2 = 6 - a$, then $a =$

A. −4　　B. −2　　C. 2　　D. 4　　E. 8

EXAMPLE 2—

$3x + 11 = 17.$

A. $6x + 22$ B. 34

EXAMPLE 3—

$H \times 2/3 = 2/3 \times 5/9, H =$

A. 5/9   B. 9/5   C. 5   D. 9   E. 14

EXAMPLE 4—

$\dfrac{2}{M} + \dfrac{2}{M} = 8.$ $M =$

A. 1/8   B. 1/4   C. 1/2   D. 2   E. 64

EXAMPLE 5—

If $5a = 6$ and $7b = 8$, $35ab =$

A. 4/3   B. 48/35   C. 35/48   D. 14   E. 48

EXAMPLE 6—

$4 \times 4 \times 4 \times 4 \times 4 = \dfrac{32 \times 32}{S}$ $S =$

EXAMPLE 7—

$5/a = 1$ $b/3 = 5$ $\dfrac{a + 3}{b + 4} =$

A. 8/19   B. 19/8   C. 2/5   D. 5/2   E. 4/9

EXAMPLE 8—

$c - 5 = d$

A. $c - 8$ B. d

EXAMPLE 9—

⅓ of a number is one more than ¼ of the number. The number is

A. 3   B. 4   C. 12   D. 36   E. 144

EXAMPLE 10—

$$\frac{8}{5} = \frac{5}{x} \qquad x =$$

A. 8/25   B. 5/8   C. 8/5   D. 25/8   E. 200

OK, LET'S SEE THE ANSWERS

EXAMPLE 1—

$a - 2 = 6 - a$

$2a = 8$ so $a = 4$.   D. Hopefully very easy for you.

EXAMPLE 2—

Do NOT solve for x. Notice $6x + 22 = 2(3x + 11)$ and $2(17) = 34$. They are equal, C.

EXAMPLE 3—

It is as easy as it looks. 5/9, A. (The commutative law for those who want to be technical.)

EXAMPLE 4—

Multiply each term by M. $2 + 2 = 8M$. $8M = 4$. $M = 4/8 = ½$.   C.

EXAMPLE 5—

Again, don't solve. $35ab = (5a)(7b) = 6(8) = 48$.   E.

EXAMPLE 6—

$$\frac{4 \times 4 \times 4 \times 4 \times 4}{1} = \frac{32 \times 32}{S}$$

Soooo $\dfrac{S}{1} = \dfrac{\overset{\;2}{\cancel{32}} \times \overset{\;2}{\cancel{32}}}{\underset{}{4 \times 4 \times 4 \times 4 \times 4}} = 4/4 = 1$.

Okay, okay, you might use a calculator here, but it still is much better if you could do this in your head.

EXAMPLE 7—

This one you have to solve. By cross multiplying,

$5/a = 1/1$ so $a = 5$. $b/3 = 5/1$.

So $b = 15$.

$$\frac{a+3}{b+4} = \frac{5+3}{15+4} = 8/19. \qquad A.$$

You must do enough of these to know the difference between Examples 5 and 7.

EXAMPLE 8—

$c - 8 = c - 5 - 3 = d - 3 < d.$ Answer is B.

EXAMPLE 9—

A reading problem.

$$\frac{1}{3}x = 1 + \frac{1}{4}x \qquad (\text{“is” is the = sign})$$

Multiply by LCD 12, every term!!!

$4x = 12 + 3x.$ $x = 12.$ C.

EXAMPLE 10—

Cross multiply. $8x = 25$, $x = 25/8$.   D.

These are very important. Let's try some more.

PROBLEMS

EXAMPLE 11—

If $2x + 9 = -14$ $x =$

A. -23 B. $-23/2$ C. $-5/2$ D. 5   E. $23/2$

EXAMPLE 12—

$x + 7 = x + -b$ $b =$

A. 7   B. −7   C. −x   D. x   E. −2x + 7

EXAMPLE 13—

$$\frac{(20 + 50) + (30 + M)}{2} = 70. \qquad M =$$

A. 30   B. 40   C. 50   D. 60   E. 70

EXAMPLE 14—

a + b is five more than a − b. Which has exactly one value?

A. a   B. b   C. a + b   D. a − b   E. ab

EXAMPLE 15—

I weigh 9 kg more than I did a year ago. My weight then was 9/10 of my weight now, how much do I weigh now?

A. 72   B. 81   C. 90   D. 99   E. 108

EXAMPLE 16—

$(8/9)y = 1$ $(4/9)y =$

A. 1/3   B. 1/2   C. 2/3   D. 3/2   E. 7/6

EXAMPLE 17—

A 50-cm piece is cut into 3 pieces: The first is 3 cm shorter than the second and the third is 4 cm shorter than the first. The length of the shortest piece is

A. 10   B. 13   C. 15   D. 17   E. 20

EXAMPLE 18—

$$b = y + \frac{1}{3} = \frac{y + 2}{3} \qquad b =$$

A. 1/2   B. 2/3   C. 3/4   D. 4/5   E. 5/6

EXAMPLE 19—

$5x - 3 = 4c.$ Then $\dfrac{5x - 3}{2} =$

A. 9/4   B. 9/2   C. c   D. 2c   E. 4c

OKAY, LET'S HAVE THE SOLUTIONS

EXAMPLE 11—

Straight algebra:

$2x = -23.$ $x = -23/2.$ B.

EXAMPLE 12—

Cross out x's from both sides because they are added. $7 = -b.$ Multiply both sides by $-1.$ $b = -7.$ B.

EXAMPLE 13—

Write 70/1 and cross multiply. $100 + M = 140,$ $M = 40.$ B.

EXAMPLE 14—

You might throw up your hands and say, "What do I do? What do I do?" Write something. Write what you read. $a + b = 5 + a - b.$ The a's cancel. $b = 5 - b.$ We could solve for b except we don't have to. The question asked is what can we solve for, not the answer. B (b) is the answer.

EXAMPLE 15—

Straight algebra, and not too easy.

$x =$ weight then. $x + 9$ is weight now.

weight then is 9/10 of weight now.

$x = 9/10(x + 9)$ Multiply by LCD, 10.

$10x = 9(x + 9)$

$10x = 9x + 81$ $x = 81$ kg.

My weight now. Well almost now. Mine is about 5 kg more as of this writing. Oh, by the way, the answer is B.

EXAMPLE 16—

Do NOT NOT NOT solve for y. You should notice 4/9 is ½ of 8/9. ½ of 1 is ½, answer is B.

EXAMPLE 17—

Let x = piece 2. $x - 3$ = piece 1. $x - 3 - 4 = x - 7$ is piece 3.

$x + x - 3 + x - 7 = 50.$ $3x - 10 = 50.$ $3x = 60.$
$x = 20.$

But this is NOT the answer.

We want the shortest piece!!!! $x - 7 = 20 - 7 = 13.$
Answer is B.

Let's continue.

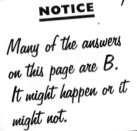

NOTICE

Many of the answers on this page are B. It might happen or it might not.

EXAMPLE 18—

Multiply by 3. $3y + 1 = y + 2.$ $2y = 1.$ $y = ½.$ $b = y + ⅓ = ½ + ⅓ = ³⁄₆ + ²⁄₆ = ⁵⁄₆.$ Answer is E.

EXAMPLE 19—

Divide by sides by 2; again we don't solve.

$$\frac{5x - 3}{2} = \frac{4c}{2} = 2c. \qquad D.$$

Enough of these already (although on the SAT most of us would like a lot of these). Let's try something else.

CHAPTER 11

EQUATIONS: TWO (OR MORE) UNKNOWNS AND QUADRATICS

When the SAT decided to allow calculators, certain problems couldn't be asked any more. Some things had to be added. Exponential questions were one, and this section was another.

You must know how to factor and solve basic quadratics.

TYPE I

Solve quadratics by factoring.

A. $x^2 - 2x - 15 = 0$. Factor.

$(x - 5)(x + 3) = 0$. Set each factor equal to zero.

$x - 5 = 0$ or $x + 3 = 0$. $x = 5, -3$.   (Last steps are better done in your head.)

B. $x^3 + x^2 = 6x$. Get everything to one side.

$x^3 + x^2 - 6x = 0$. Factor. It's a cubic, degree 3, 3 answers.

$x(x + 3)(x - 2) = 0$ $x = 0, -3, 2$

TYPE 2

Solve for x by square rooting.

A. $x^2 = 49$ Take $\pm$ square roots of both sides.

   $x = \pm\sqrt{49}$ $x = +7$ or $x = -7$.

B. $5x^2 = 45$.   Divide by 5.

   $x^2 = 9$ $x^2 = \pm3$

NOTE: This also can be solved by factoring: $x^2 - 49 = 0$, . . . but better by square rooting. Hooray for the square!

TYPE 3

Solving two equations in two unknowns.

A. By addition

   $x + y = 28$

   $x - y = 4$ Add together.

   $2x + 0 = 32$ $x = 16$ Substitute in Eq. (1) or (2).

   $16 + y = 28$ $y = 12$

B. By subtraction

   $4x + y = 18$

   $x + y = 6$ Subtract.

   $3x \phantom{{}+ y} = 12$ $x = 4$ Sooo $y = 2$. From Eq. (2).

I know, I know. As many of you know, there are many more types than these listed. But up to this point, the SAT has only asked these. My books *Algebra for the Clueless* and *Precalc with Trig for the Clueless* may have more than you need or want.

LET'S DO SOME PROBLEMS!!!!

EXAMPLE 1

$x + y = 5$ $x - y = 9$

A. x B. y

EXAMPLE 2—

$2x + y = 10$ $x + 2y = 2$

A. $x + y$ B. 4

EXAMPLE 3—

$(x + 7)(x - 5) = 0$

A. x B. 7

EXAMPLE 4—

$x^2 + 11x + 14 = 0$

A. $x^2 + 11x$ B. 14

EXAMPLE 5—

$x^2 = 4$, then $x^3 =$

A. 6   B. 8 only   C. −8 only   D. 6   E. 8 or −8

EXAMPLE 6—

The sum of 2 numbers is 26. The difference is 12.

A. larger of the 2 integers   B. 20

EXAMPLE 7—

$(x + 2)(1/x) = 0$ $x =$

A. 2   B. 0   C. −2   D. −2 or 0   E. any integer

EXAMPLE 8—

I buy a table tennis racket and one can of balls for $42.00. My friend buys the same racket and two cans of balls for $45. How much does a racket cost?

A. $39   B. $40   C. $41   D. $42   E. $45

EXAMPLE 9—

$2a + 3b = 34$, $a + 2b = 14$, $\dfrac{3a + 5b}{2} =$

A. 20   B. 24   C. 36   D. 40   E. 48

LET'S CHECK SOME OF THESE

EXAMPLE 1—

Adding $2x = 14$. $x = 7$. $y = -2$.   Answer is A.

EXAMPLE 2—

Adding, we get $3x + 3y = 3(x + y) = 12$. Divide by 3. $x + y = 4$. They are equal!!!   C.

EXAMPLE 3—

$x = -7, 5$ (opposite signs). 7 is bigger than both.   B.

EXAMPLE 4—

Phoney phactoring or foney factoring. $x^2 + 11x$ must $= -14$, since $-14 + 14 = 0$.   Answer is B.

EXAMPLE 5—

$x = \pm 2$. Soooo $(\pm 2)^3 = +8$ or -8.   Answer is EEEEE.

EXAMPLE 6—

Straightforward. $x + y = 26$. $x - y = 12$. $2x = 38$. $x = 19$. $y = 7$ (not important).   B.

EXAMPLE 7—

$1/x$ is never $= 0$ since x is in the bottom. Answer is $x = -2$.   C.

EXAMPLE 8—

$r + 2b = 45$
$r + \ b = 42$

Subtracting, the balls $b = 3$; racket $r = 39$.   A.

EXAMPLE 9—

Adding, we get $3a + 5b = 48$. So $(3a + 5b)/2 = 24$.   B.

These are fun or phun.

LET'S DO SOME MORE!!!

EXAMPLE 10—

$x + y = 11$. Then $2x + 2y =$

A. 13   B. 11/2   C. 22   D. 44   E. 88

EXAMPLE 11—

$x + 2y = 12$. $2x - 2y = 6$.

A. x   B. y

EXAMPLE 12—

$b^2 = c^2$

A. bc   B. c^2

EXAMPLE 13—

What are all solutions to the equation $x^2 - 4x = 0$?

A. 0   B. 4   C. −4   D. 0, 4   E. 0, 4, −4

EXAMPLE 14—

$\dfrac{x}{6} = x^2$.   I. −1/6   II. 0   III. 1/6

A. I only   B. II only   C. III only   D. II and III
only   E. I, II, and III

EXAMPLE 15—

In a class of 60 students, the number of boys is twice
the number of girls. Which of the following accurately
describes the situation?

A. b + 2g = 60   B. b − 2g = 0   C. 2b − g = 0   D. b + g = 60   E. b + 2g = 60
   b − g = 60   b + g = 60   b + g = 60   b + 2g = 0   b − 2g = 0

EXAMPLE 16—

If $m^2 = 16$ and $n^2 = 36$, the difference between the largest value of m − n and the smallest value of m − n is . . .

A. 20   B. 10   C. 4   D. 2   E. −2

EXAMPLE 17—

$5m + 4n = 14$ and $2m + 3n = 14$. $7(m + n) =$

A. 7   B. 14   C. 21   D. 28   E. 35

EXAMPLE 18—

2a / a

The area is 32. a =

A. 2   B. 4   C. 6   D. 8   E. 16

NOW THE ANSWERS

EXAMPLE 10—

$x + y = 11$. $2x + 2y = 2(x + y) = 2(11) = 22$.   C.

EXAMPLE 11—

Adding $3x = 18$. $x = 6$. So $y = 3$.   A.

EXAMPLE 12—

Tricky. If, let's say, b and c both = 6 or −6, they are equal. If b = 6, c = −6, the squares are both 36, but B is clearly bigger. Because we can't always tell which is bigger, the answer is DDDD.

EXAMPLE 13—

Factoring, we get $x(x − 4) = 0$ sooo x = 0 or 4.   D.

EXAMPLE 14—

Not so easy. You must solve and ignore choices. Multiply by 6 and get everything to one side.

You can also do this by substitution, but solving is probably a little better.

$6x^2 − x = 0$. $x(6x − 1) = 0$. $x = 0$ or $6x − 1 = 0$. $6x = 1$. $x = 1/6$.   Answer is D.

EXAMPLE 15—

Reading. You must be able to read.

$b + g = 60$. $b = 2g$ or $b - 2g = 0$.   Answer is B.

With calculators, there will be more of these. I know it sounds cruel, but. . . .

EXAMPLE 16—

$m^2 = 16$ and $n^2 = 36$. $m = \pm 4$ and $n = \pm 6$.

Max of $m - n$ is the largest m and smallest n. $m - n = 4 - (-6) = 10$.

Min of $m - n$ is smallest m and the largest n. $m - n = -4 - 6 = -10$.

The difference between the largest and the smallest is $10 - (-10) = 20$.   A.

I think this is a tough one, especially when you must do it and read it quickly. You must be careful.

EXAMPLE 17—

Adding, we get $7m + 7n = 7(m + n) = 28$. That's it.   D.

EXAMPLE 18—

Area of a rectangle is base times height.

$2a^2 = 32$ $a^2 = 16$ $a = 4$ B.

Hey!!!! How did geometry sneak in here???? Sometimes the questions cross over two or more topics. Besides, we've done enough of these. Believe it or not, we've basically finished the algebra. Let's do some angles, then more on areas. Then a little more. And awaaaay we goooo!!!!

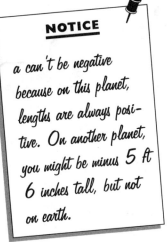

NOTICE

a can't be negative because on this planet, lengths are always positive. On another planet, you might be minus 5 ft 6 inches tall, but not on earth.

CHAPTER 12

GEOMETRY: FIGURE THE ANGLE

In doing an unofficial survey of about 40 SATs, out of 60 questions, an average of 12 or 20% of the math is geometry. Sometimes it is as low as 15% and sometimes as high as 25%. Therefore, geometry is verrry important. However, THERE ARE NO PROOFS. Here are some of the facts you need to know. . . .

AB perpendicular to CD (notation AB CD) means 90° angle. ∡ADC and ∡CDB are complementary angles because they add up to 90°. (Notice the spelling of *complementary*. A compliment is how beautiful or handsome you are.)

1.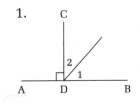

2.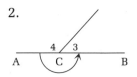

Angle ACB is 180°, half way around the circle. ∡3 and ∡4 are supplements because they add to 180°.

3.
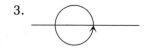

Once around a circle is 360°.

4.
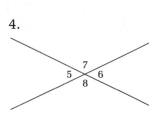

∡5 and ∡6 are vertical angles. They are equal. (I know, I know, in most of your geometry classes, they are congruent, but not when I went to school. I am rebelling backward.) So are ∡7 and ∡8. Notice ∡5 and ∡7 are supplements. So are ∡6 and ∡7, ∡6 and ∡8, and ∡8 and ∡5.

5.
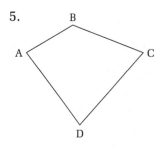

The sum of the angles of any quadrilateral (4-sided figure) is alllllways 360°.

6.
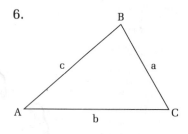

The sum of the angles of any triangle is . . . 180°.

Capital letters indicate vertices (points).

The side opposite A is little a.

The largest side lies opposite the largest angle.

So, if the figure were drawn to scale,

and a < c < b, thennnn

∡A < ∡C < ∡B.

Angle x is called an EXTERIOR ANGLE, formed by one side of a triangle annnd one side of the triangle extended.

∡1 + ∡2 = ∡x

7.

8.

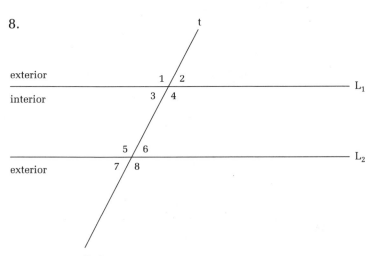

L$_1$ is parallel to L$_2$.

t = transversal, any line that cuts two or more lines.

∡4 and ∡5, ∡3 and ∡6 are called alternate interior angles and are EQUAL.

∡1 and ∡8, ∡2 and ∡7 are called alternate exterior angles and are EQUAL.

∡2 and ∡6 (upper right), ∡4 and ∡8 (lower right), ∡1 and ∡5 (upper left), and

∡3 and ∡7 (lower left) are called corresponding angles and are EQUAL.

∡3 and ∡5, and ∡4 and ∡6 are called interior angles on the same side of the transversal and are SUPPLEMENTARY (add to 180°).

8 revisited. Suppose you don't want to learn all of this. YOU DON'T HAVE TO.

All the angles less than 90°, ($\angle 2$, $\angle 3$, $\angle 6$, $\angle 7$) are equal.

All the angles more than 90°, ($\angle 1$, $\angle 4$, $\angle 5$, $\angle 8$) are equal.

Any two angles not equal must add to 180°. That's it!!!!

LET'S DO SOME PROBLEMS

EXAMPLE 1

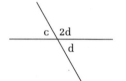

c =

A. 30°   B. 60°   C. 90°   D. 120°   E. 180°

EXAMPLE 2

$L_1 \parallel L_2$. x + y =

EXAMPLE 3

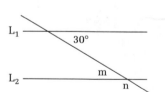

$L_1 \parallel L_2$. n − m =

A. 150°   B. 120°   C. 90°   D. 60°   E. 30°

EXAMPLE 4

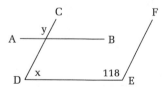

AB ∥ DE annnnd CD ∥ EF. x + y =

A. 180°   B. 150°   C. 120°   D. 90°
E. can't be determined from the information given

EXAMPLE 5—

b − a =

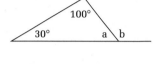

EXAMPLE 6—

$L_1 \| L_2$. b =

A. 30°   B. 40°   C. 70°   D. 80°   E. can't be determined from the information given.

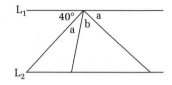

EXAMPLE 7—

y =

A. 40°   B. 42°   C. 50°   D. 58°   E. can't be determined from the info given

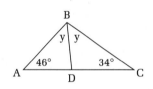

EXAMPLE 8—

b + c =

A. a/2 − 180°   B. 180° − a/2   C. 180 − 5a/2
D. 180 − a   E. a − 180°

EXAMPLE 9—

A. 2y°   B. y + 20°

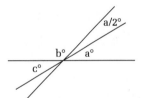

EXAMPLE 10—

(figure not drawn to scale)

y − 20° =

A. 40°   B. 50°   C. 60°   D. 70°   E. 80°

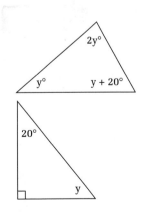

EXAMPLE 11—

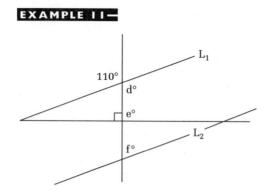

$L_1 \parallel L_2$. $d + e + f =$

A. 180°   B. 220°   C. 230°   D. 270°   E. 330°

EXAMPLE 12—

$DE \parallel AC$ $a =$

A. 70°   B. 50°   C. 40°   D. 30°   E. 20°

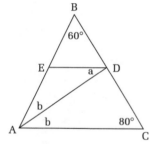

HOW ABOUT SOME ANSWERS???? OK!

EXAMPLE 1—

$2d + d = 180°$ (supplements). $d = 60° = c$ (vertical angles).   B.

EXAMPLE 2—

x and y are clearly not equal. They must add to 180°.

EXAMPLE 3—

Angle $m = 30°$ (alternate interior). $n = 150°$ (sup). $n - m = 120°$.   B.

EXAMPLE 4—

We only need to know that AB and DE are parallel, and the two angles are not equal. They must add (like Example 2) to 180°.   A.

NOTE

If it says "cannot be determined," sometimes it can and sometimes it can't. It has been my observation that if it says "cannot be determined" twice on one SAT, both won't be "can't be determined," at least so far.

EXAMPLE 5—

30 + 100 + a = 180. a = 50. a and b are sups. b = 130°.
b − a = 80°.

EXAMPLE 6—

As much as you try, you can't find b. E. This may be the toughest of all, because you must make a quick decision it can't be determined.

EXAMPLE 7—

Easier than it looks. In the big triangle, 46 + 34 + 2y = 180. 2y = 100. y = 50.   C C C!!!

EXAMPLE 8—

Look at the dark line. a/2 + b + c = 180. Soooo b + c = 180 − a/2.   B.

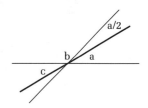

EXAMPLE 9—

y + 2y + y + 20 = 180. 4y = 160. y = 40.

So 2y = 80 and y + 20 = 60. So A is bigger.

EXAMPLE 10—

y is 70° (90 − 20). y − 20 is 50°. This doesn't look too hard if you have plenty of time. However, if you move quickly, as you must, you must READ this accurately.

NOTE

"Figure not drawn to scale" usually means two things:

1. In a simple figure, it means not drawn to scale.

2. In a complicated figure, it probably is drawn to scale even though it says not.

EXAMPLE 11—

d = 110° (vertical). e = 90° (sup of a rt angle). L_1 parallel to L_2. 110 and f not equal. They are sups. f = 70°. d + e + f = 110 + 90 + 70 = 270°.   D.

EXAMPLE 12—

Bigggg triangle ABC, 2b + 60 + 80 = 180. 2b = 40. b = 20. b = a (alternate interior).   E.

The first time I stopped here, I was tired. When I edited this for the first time, I ached from painting. Looks like this is the quitting page for me, but not for you. I'll pick this up tomorrow with more problems. But I really like these. My body doesn't seem to.

PROBLEMS

EXAMPLE 13—

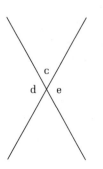

d + e = 3c

A. (3/2)c   B. d

EXAMPLE 14—

CD bisects ⵥACB. x =

A. 35°
B. 40°
C. 45°
D. 50°
E. 70°

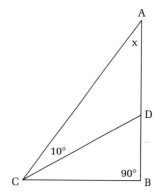

EXAMPLE 15—

ⵥABC is 11x degrees. What can the value of x be?

A. 12   B. 8
C. 4   D. 2   E. 1

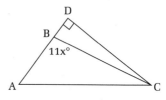

EXAMPLE 16—

b = 3a   c =

A. 30°
B. 60°
C. 70°
D. 80°
E. 90°

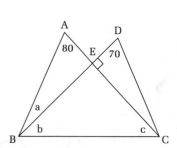

EXAMPLE 17—

A. p + q + r   B. s + t + u

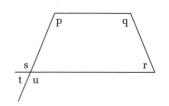

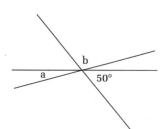

EXAMPLE 18—

b in terms of a is

A. 130 − a
B. 230 − a
C. 260 − 2a
D. 50 + a

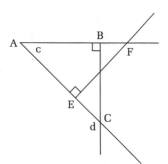

EXAMPLE 19—

d in terms of c

A. 90 + c
B. 90 + 2c
C. 180 − c
D. 180 − 2c
E. 2c

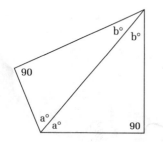

EXAMPLE 20—

a + b =

A. 30
B. 45
C. 60
D. 90
E. 180

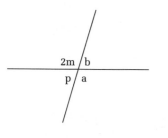

EXAMPLE 21—

180 − m =

A. m + a
B. m + p
C. a + p
D. a + b
E. p + b

EXAMPLE 22—

$x + y + z =$

A. 180
B. 210
C. 240
D. 270
E. 360

EXAMPLE 23—

$w =$

A. 10°
B. 12½°
C. 16°
D. 20°
E. Cannot be determined from info given

EXAMPLE 24—

$y =$

A. 30°
B. 33°
C. 36°
D. 40°
E. 45°

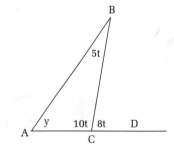

EXAMPLE 25—

A triangle has angles C, D, 100°

A. 90°   B. C

OK, LET'S SOLVE THEM

EXAMPLE 13—

Since d = e, verticals, d + e = 2d = 3c. d = 3c/2 (dividing by 2). C, equal.

EXAMPLE 14—

Angle ACB is 20° (double 10°). x = 70° since the sum of the angles of a triangle is 180°.

EXAMPLE 15—

Angle ABC is the exterior angle of triangle BDC and must be more than either interior angle, more than 90°. Only when x = 12, (12)(10) = 120°, can this happen.   A.

EXAMPLE 16—

This can be done quicker than it can be written. In triangle DEC, missing angle ECD is 20°, which you should probably write on your test paper. Angle AEB vertically equals 90°. In triangle ABE, missing angle, a = 10°. b = 3a = 30°. Angle BEC is sup of 90°, which is 90°. b = 30°. Therefore, c = 60°.   B. Hmmmm. Maybe not quite trivial.

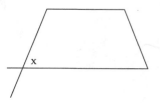

EXAMPLE 17—

Labeling missing angle x, s + t + u + x = 360. p + q + r + x also = 360. s + t + u + x = p + q + r + x. Subtracting x, we see they are = C.

EXAMPLE 18—

b + vertical of a (= a) + 50 = 180. Sooo b = 130 − a.   A.

EXAMPLE 19—

Here is a typical example of what the SAT does. Line EF is there only to fool you. d is the exterior angle of triangle ABC. Focus on it. d = 90 + c.   A.

EXAMPLE 20—

$2a + 2b + 90 + 90 = 360.$ $2a + 2b = 180.$
$a + b = 90.$ D.

EXAMPLE 21—

$p + 2m = 180°.$ $p + m + m = 180.$ $p + m =$
$180 - m.$ B.

EXAMPLE 22—

The sum of the nine angles of 3 triangles is 540. $540 -$
4 rt angles, $540 - 360 = 180 = x + y + z.$ A.

EXAMPLE 23—

Verrrry verrrry tricky. Lots of verticals. None should be
used.

On top, $5x + 3x + 20 = 180.$ $8x = 160.$ $x = 20.$
$3x + 20 = 8w.$ $3(20) + 20 = 80 = 8w.$ $w = 10.$ A.

EXAMPLE 24—

Sups $10t + 8t = 180.$ $18t = 180.$ $t = 10.$ $10t = 100.$ $5t = 50.$
$y = 30°.$ A.

EXAMPLE 25—

The sum of the angles of a triangle is 180°. If one is
100, the other two must be less than 80. So A is larger.

Most of these aren't too tough, and you must learn
them because geometry is a substantial part of the SAT.
But there is more.

This is a kind of problem that shows up every once in
a while. When it does, it is always at the end of a test,
one that if you worked out the "right" way, would take
too long. It is the sum of "a lot of angles." I can't
describe it any other way. All you need to know is the
following (in triangles only):

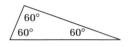

Let all interior angles = 60°, no matter what size.

All exterior angles = 120°.

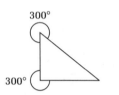

All "outside" angles = 300°.

Why will this work, especially when you know it's not true sometimes? Because if it is always true, it is true in the specific case when all the angles are equal. Why should you do it this way? The right ways will be too slow. What if I want to know the right way? Write me. I'll answer.

PROBLEMS

EXAMPLE 1—

$x + y + z - (a + b + c)$

EXAMPLE 2—

The sum of the 6 angles is

EXAMPLE 3—

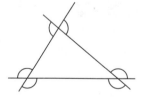

The sum of the 6 angles is

EXAMPLE 4—

The sum of the 3 marked angles is

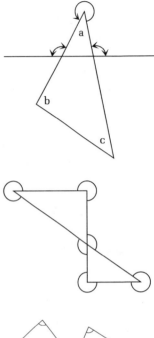

EXAMPLE 5—

The sum of the 6 marked angles is

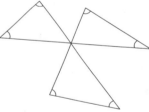

EXAMPLE 6—

The sum of the 6 marked angles is

SOLUTIONS

The trick solutions are very short. The real ones are not.

EXAMPLE 1—

$3(120) - 3(60) = 180°$.

EXAMPLE 2—

$6(120) = 720$.

EXAMPLE 3—

$6(60) = 360$.

EXAMPLE 4—

$2(120) + 1(300) = 540$. The letters are there to fool you.

EXAMPLE 5—

$4(300) + 2(120) = 1440$.

EXAMPLE 6—

$6(60) = 360$.

They are always at the end of a section. With the trick, this is one right answer you probably would not get. But these problems rarely show up. Let's get back to more common ones.

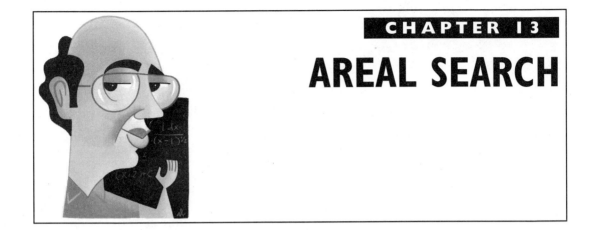

CHAPTER 13

AREAL SEARCH

The SAT usually asks a number of questions on area.
Here are five formulas you should know, and two more
it would be nice to know.

Rectangle

$A = b \times h$

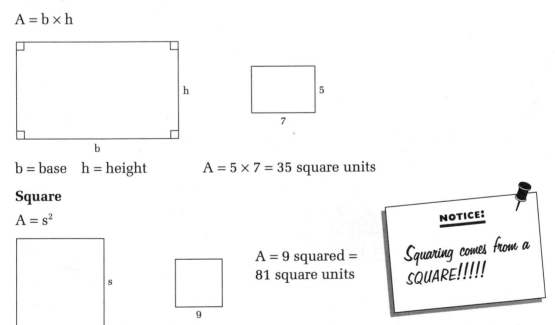

b = base   h = height   A = 5 × 7 = 35 square units

Square

$A = s^2$

A = 9 squared =
81 square units

NOTICE:

*Squaring comes from a
SQUARE!!!!!*

Circle

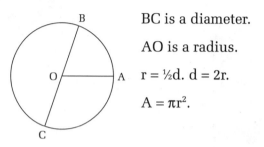

BC is a diameter.

AO is a radius.

$r = \frac{1}{2}d$. $d = 2r$.

$A = \pi r^2$.

Diameter: line segment from one side of the circle to the other through the center

Radius: line segment from the center to the outside, π = approx 3.14

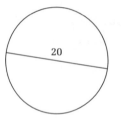

$d = 20$

sooo $r = 10$

$A = \pi r^2 = \pi 10^2 = 100\pi$

Answer on the SAT almost always in terms of π.

Triangle

$A = \frac{1}{2}b \times h$

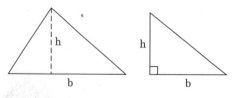

$A = \frac{1}{2} \times 9 \times 7 = 63/2 = 31.5$ square units.

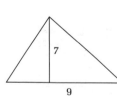

Sector

The shaded figure is called a *sector*. It is part of a circle. It usually is skipped, but it shouldn't be.

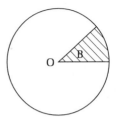

Center is O. What part of a circle, you ask? B/360.

So, the area of a sector is $(B/360) \times \pi r^2$.

$A = (40/360)\pi(6)^2 = 4\pi$ square units.

That's all there is, but it is important!

Trapezoid (nice to know because it has shown up)

$A = \frac{1}{2}h(b_1 + b_2)$
$h = $ height
$b_1 = $ upper base
$b_2 = $ lower base

Bases are parallel.
Other sides are not.
Bases are never equal.
Other sides might be.

$A = \frac{1}{2}h(b_1 + b_2) = \frac{1}{2}(8)(7 + 13) = 80$ square units

You really don't need to know this, because trapezoids can be broken up into triangles and rectangles. It is nice to know.

Equilateral triangle

$$A = \frac{s^2\sqrt{3}}{4}$$

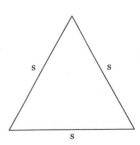

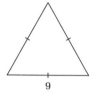

9

$$A = 9^2 \frac{\sqrt{3}}{4} = 81 \frac{\sqrt{3}}{4}$$

It's really important you know the areas. Sometimes the SAT asks more than one. Let's try some problems. Some are realllly innnnteresting.

PROBLEMS

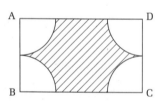

EXAMPLE 1—

ABCD is a square.

A. $16b^2$ B. $25c^2$

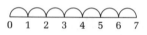

EXAMPLE 2—

The area of the 7 semicircles is

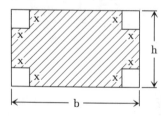

EXAMPLE 3—

Rectangle ABCD, 4 arcs center at A, B, C, D. AD = 12. AB = 6. Area of shaded portion is

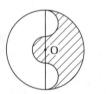

EXAMPLE 4—

Circle center O. 3 equal semicircles. Area of large circle is 36 π. Area of the shaded portion is

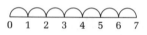

EXAMPLE 5—

Area of the shaded portion is. . . .

EXAMPLE 6—

The length of rectangle C is 20% longer than rectangle D. The width of rectangle C is 20% less than rectangle D. The area of rectangle C is

A. 20% greater than rectangle D

B. 4% greater than rectangle D

C. Equal to rectangle D

D. 4% less than rectangle D

E. 20% less than rectangle D

EXAMPLE 7—

Area of triangle ABC is 40. Area of triangle BCD is

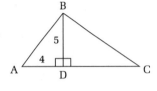

EXAMPLE 8—

ABCD is a square. Area is

A. 9   B. 4   C. 1   D. ¼   E. cannot be determined from the info given

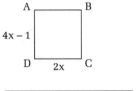

EXAMPLE 9—

The area of the shaded portion is

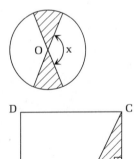

EXAMPLE 10—

Center of the circle is O. Radius of the circle is 8. The total area of the shaded portion is 16π. x =

A. 45   B. 90   C. 120   D. 135   E. 270

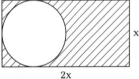

EXAMPLE 11—

XB = ¼AB. Area of triangle CBX is 10. The area of rectangle ABCD is. . . .

LET'S LOOK AT SOME ANSWERS

EXAMPLE 1—

Because it is a square, 4b = 5c. $(4b)^2 = (5c)^2$. $16b^2 = 25c^2$. They are equal.   C.

EXAMPLE 2—

Kinda tricky. Diameter of each semicircle is 1. Sooo, radius is ½. 3½ or 7/2 circles. Area is $7/2(\pi(½)^2) = (7/8)\pi$.

EXAMPLE 3—

Area is 1 rectangle minus 4 (¼ circle) or 1 rectangle minus 1 circle, radius 3.

$A = (12)(6) - \pi(3)^2 = 72 - 9\pi$

EXAMPLE 4—

Picture is this.   You should see this.

½ large semicircle − ½ small semicircle.

Large circle is 36π. Radius is 6.

Area of semicircle is 18π.

Radius of smaller circle is $(1/3)6 = 2$.

Area of smaller circle is 4π.

Area of smaller semicircle is 2π.

$18\pi - 2\pi = 16\pi$

EXAMPLE 5—

This is a rectangle with 4 squares cut out. $bh - 4x^2$.

EXAMPLE 6—

Say rectangle D is 10 by 10 (remember a square is a rectangle). Rectangle C would be 8 by 12 (20% longer by 20% shorter). 96, which is 4% less.   D.

EXAMPLE 7—

Area of triangle ABD is ½(5)(4) = 10. Big triangle is 40. Right most triangle is 40 − 10 = 30. You don't have to do anything else.

EXAMPLE 8—

Square means $4x - 1 = 2x$. $2x = 1$. We don't have to solve for x since 2x is the side of the square!! $1^2 = 1$! The answer is . . . C, C, C.

EXAMPLE 9—

Shaded portion issss rectangle minus circle whose diameter is x. Radius is x/2.

$A = bh - \pi r^2 = x(2x) - \pi(x/2)^2 = 2x^2 - \pi x^2/4$

EXAMPLE 10—

Area of the circle is $\pi r^2 = 64\pi$. The shaded part, 16π, is ¼ of circle.

Unshaded part is 3/4 of circle.

3/4 of 360° = 270°   2x = 270°   x = 135°   D.

EXAMPLE 11—

Really tricky and much easier than it looks. We can't find base or height, but we don't need to.

8 triangles make up the rectangle, which hopefully you can see without drawing. $8 \times 10 = 80$.

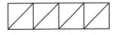

LET'S TRY SOME MORE. . . . .

EXAMPLE 12—

Area of the square with 2 sides tangent to the circle is $4b^2\pi$. The area of the circle is

A. $b\pi^2$ B. $b^2\pi^2$ C. $b^2\pi$ D. $2b^2\pi$ E. can't be determined

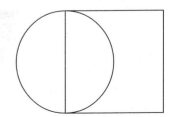

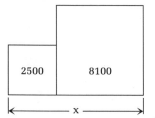

EXAMPLE 13—

Areas are given. Both are squares. x =

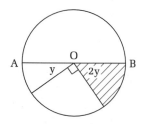

EXAMPLE 14—

Center of the circle is O. Radius is 10. Area of shaded portion is 80π. x =

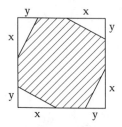

EXAMPLE 15—

Center is O. AB = 2. AB diameter. Area of shaded sector is

A. $\pi/12$ B. $\pi/8$ C. $\pi/6$ D. $\pi/4$ E. $\pi/3$

EXAMPLE 16—

The sum of the areas of two squares if the sides are 2 and 3, respectively. . . .

A. 5   B. 6   C. 10   D. 13   E. 25

EXAMPLE 17—

The square has area of $4x^2$. If a rectangle with width x has the same area as the shaded portion, the length would be

A. $x - y$ B. $2x - y$ C. $x - 2y$ D. $4x - 2y$
E. $2x - 4y$

EXAMPLE 18—

The area of the rectangle and triangle are the same.
cd/2 = 80. ab =

A. 320   B. 160   C. 80   D. 40   E. 20

EXAMPLE 19—

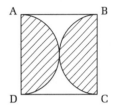

ABCD is a square, and the two shaded areas are semi-circles. AB = 20. Find the ratio of the shaded region to the ratio of the area of the square.

EXAMPLE 20—

Two circles have radius R + 2 and R. The difference in areas is 8π. The larger radius is

A. 1   B. 2   C. 3   D. 4   E. 9

EXAMPLE 21—

The area of the rectangle is 1. y =

A. 7/22   B. 22/7   C. 3/7   D. 7/3   E. 7

LET'S LOOK AT THE ANSWERS

EXAMPLE 12—

This is a very difficult one. $s^2 = 4b^2\pi$. Sooooo $s = \sqrt{\pi 4b^2} = 2b\sqrt{\pi}$. d = s. $r = \frac{1}{2}s = b\sqrt{\pi}$. $A = \pi r^2 = \pi(b\sqrt{\pi})(b\sqrt{\pi}) = b^2\pi^2$.   B.

EXAMPLE 13—

A breather. Area of one square is 8100. Side is 8100 = 90. 2500 is 50. 50 + 90 = 140, which is the value of x.

EXAMPLE 14—

Another toughie. Circle radius 10 has area $\pi r^2 = 100\pi$. 80π/100π is 4/5 of the circle. The unshaded part of the circle is 1/5 of the circle and 1/5 of 360° is 72°. Angle x is 90 − 72 = 18°. Whew!!!!

EXAMPLE 15—

AO is 1. Area is $\pi r^2 = \pi(1)^2 = \pi$. y + 2y + 90 = 180°. 3y = 90. y = 30°. 2y = 60°. 60°/360° = 1/6 of circle. The shaded portion is $(1/6)\pi$. C is the answer.

EXAMPLE 16—

Another breather. Most of the SAT geometry is easy, but I've given you a larger example of the harder types. Getting back to this problem, $2^2 + 3^2 = 4 + 9 = 13$.   D.

EXAMPLE 17—

The square = shaded plus 4 triangles. Shaded = $4x^2 - 4(\frac{1}{2})(x)(y)$. $4x^2 - 2xy$. Only on the SAT would you only factor out an x. $4x^2 - 2xy = x(4x - 2y)$.

If the width is x, the length must be 4x − 2y.   D.

EXAMPLE 18—

Super tricky. cd/2 = 80. cd = 160 = area of rectangle = area of triangle = ½ab. So ab = 2(160) = 320.   A.

EXAMPLE 19—

Area of square = $20^2 = 400$. AB = 20 = diameter of circle (2 half circles). r = 10. Area is $\pi r^2 = \pi(10)^2 = 100\pi$. Ratio is $100\pi/400 = \pi/4$.

EXAMPLE 20—

$\pi(R + 2)^2 - \pi R^2 = \pi(R^2 + 4R + 4 - R^2) = 8\pi$ 4R + 4 = 8
4R = 4   R = 1   R + 2 = 3   C.

EXAMPLE 21—

$3\frac{1}{7}$ y = (22/7)y = 1.   y = 1/(22/7) = 7/22.   A.

Enough of these; let's do perimeters!

When I went through this book for the first time, for some reason I left out all the perimeter formulas. Maybe that was because if it is anything other than a circle, the perimeter means merely add up all the outside lengths. But let me list the formulas for those who absolutely must have formulas for everything.

Perimeter is 2b + 2h.
If you forget, add up all the sides.

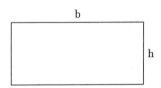

Perimeter is 4s.
If you forget, add up all the sides.

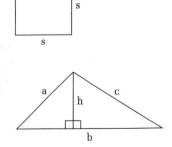

Perimeter is a + b + c. h is NOT on the outside.

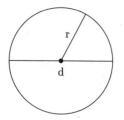

Perimeter of a circle is the circumference.
$c = \pi d = 2\pi r$.

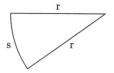

Perimeter is $2r + s$.
r = radius; s = arc length. The figure is a sector.

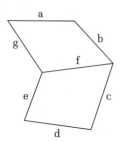

Perimeter is $a + b + c + d + e + g$.
f is not part of the perimeter.

That's all there is to it.

LET'S TRY SOME PROBLEMS

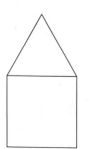

EXAMPLE 1

This is an equilateral triangle with a square of area 100 underneath. The perimeter of the figure is

A. 30   B. 40   C. 50   D. 60   E. 300

EXAMPLE 2—

A square is divided into 4 smaller squares. The perimeter of the larger square is 2. The perimeter of the smaller square is

A. 1/8   B. 1/6   C. 1/4   D. 1/2   E. 1

EXAMPLE 3—

If the perimeter of square C is triple the perimeter of square D, the area of square C is how many times the area of square D?

A. 1/3   B. 1   C. 3   D. 9   E. 27

EXAMPLE 4—

X, O, Y, Z are the midpoints of the diameters of the 4 semicircles. CD is a line segment containing the diameters of the semicircles. If CD = 12, what is the length of the dotted line from C to D?

A. 6π   B. 12π   C. 18π   D. 24π   E. 72π

EXAMPLE 5—

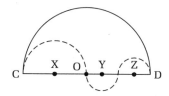

WX and YZ are the diameters, length 10. Find the lengths of the darkened arcs.

EXAMPLE 6—

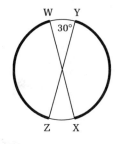

A circle has circumference 1. The radius is

A. 2π   B. π   C. 1   D. 1/π   E. 1/2π

OKAY, LET'S SOLVE THESE

EXAMPLE 1—

$A = s^2 = 100$. $s = 10$. There are 5 (not 6) 10s on the outside. $5 \times 10 = 50$.   C.

EXAMPLE 2—

If the ratio of the sides of a square (or any other figure) is ½, so are their perimeters. ½ of 2 is 1. Answer is E.

EXAMPLE 3—

Suppose one square has a side s; its perimeter is 4s. A second square's side is 3s. Its perimeter is 12s. $12s/4s = 3/1$ (triple). The area of square 1 is s^2. The area of square 2 is $(3s)^2 = 9s^2$, 9 times the amount. D. (You could also use numbers, but make them verrrry simple.)

EXAMPLE 4—

The dotted line consists of one semicircle (medium semicircle) and 1 full circle (2 halves of the itsy-bitsy circle). If CD = 12, CO = 6, diameter of medium circle ½c = ½πd = ½π6 = 3π. Itsy-bitsy circle has diameter 3. c = πd = 3π. 3π + 3π = 6π.   A.

EXAMPLE 5—

Diameter of circle is 10. c = 10π. 30° + its vertical angle 30° = 60°. The darkened arc is 360° − 60° = 300°. 300/360 = 5/6 of circle. 5/6 of 10π = 25/3π.

EXAMPLE 6—

$$2\pi r = 1 \qquad \frac{2\pi r}{2\pi} = \frac{1}{2\pi}$$

So r = ½π.   E.

LET'S TRY SIX MORE. . . .

EXAMPLE 7—

The perimeter is

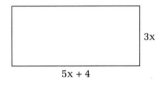

3x

5x + 4

EXAMPLE 8—

A fence is 60 feet long in front of a building. There is a fence post every 6 feet.

A. number of fence posts   B. 10

EXAMPLE 9—

A circle has center C. Area is 144π. The circle is divided into 6 equal parts. The perimeter of the shaded part issss

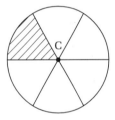

A. $24(1 + \pi)$ B. $12(2 + \pi)$ C. $4(\pi + 6)$ D. $4(\pi + 3)$
E. $12(1 + 2\pi)$

EXAMPLE 10—

If the perimeter of a 7-sided figure is 15, and each side is lengthened by 3 units, the perimeter of the new figure is?

EXAMPLE 11—

The perimeter of a rectangle is 100. What could one of the values of the sides be?

I. 45   II. 50   III. 55

A. I only   B. II only   C. III only   D. I and II
E. II and III

EXAMPLE 12—

What is half the perimeter of a square if the area is 64?

LET'S CHECK OUT THE RESULTS

EXAMPLE 7—

This is a straight perimeter problem.

$5x + 4 + 5x + 4 + 3x + 3x = 16x + 8$.

EXAMPLE 8—

A trick question. 60 divided by 6 is 10, buuuut you need a fence post at the beginning. So we need 11 fence posts, A.

EXAMPLE 9—

Semicomplicated, but it shouldn't be too hard.

$A = \pi r^2 = 144\pi$ $r^2 = 144$ $r = 12$
c (of whole circle) is $2\pi r = 2\pi(12) = 24\pi$

This sector is 1/6 of a circle. Soooo

$arc = (1/6)24\pi = 4\pi$

Perimeter is $4\pi + 2$ radii $= 4\pi + 24 = 4(\pi + 6)$. C is the one, is the one, is the one.

EXAMPLE 10—

You do not have to know the length of each side. If 3 units are added to each side, $3 \times 7 = 21$. $21 + 15 = 36$.

EXAMPLE 11—

If one side is 45, then two sides would add to 90; the other sides would be 5 and 5. It is possible. If one side is 50, $50 + 50 = 100$. Other sides would be 0 and 0, not possible. $55 + 55 = 110$, clearly not possible. A is the answer.

EXAMPLE 12—

If you had a lot of time, this problem would be very
easy. But when time is short, you must read quickly
and accurately.

$s^2 = 64$ $s = 8$ ½ the perimeter, 2s, is 16

LET'S DO A FEW MORE

EXAMPLE 13—

A. circumference of circle with diameter XY
B. perimeter of rectangle WXYZ

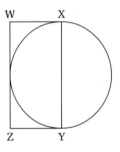

EXAMPLE 14—

Circle has diameter 2.

A. perimeter of LMNP   B. 8

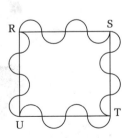

EXAMPLE 15—

RSTU is a rectangle with 16 semicircles as shown. The
total arc length is 16π.

1. What is the area of rectangle RSTU?

2. For fun, what is the area inside the curved region
    bounded by all the semicircles from R to S to T to
    U and back to R?

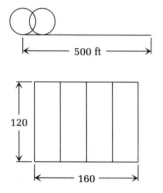

500 ft

120

160

EXAMPLE 16—

A circle of radius 10 rolls on a 500-ft line. How many revolutions does it make?

EXAMPLE 17—

If each line segment represents a fence on a farm that is divided into 4 equal regions, find the total length of the fencing needed.

LET'S CHECK THESE NOW

EXAMPLE 13—

Let's say the radius of the circle is 1 unit. WX = 1, XY = 2. The perimeter of the rectangle would be 1 + 2 + 1 + 2 = 6 units. radius = 1. C = 2πr = 2π(1) = 2π = 2(3.14) > 6. Answer is A.

EXAMPLE 14—

The diameter of the circle is 2. So 4 of them would be 8. At least one of LM, MN, NP, PL is less than the diameter (actually more than one). Because the diameter is the largest chord of a circle, the perimeter of the figure must be less than 8. Answer is B.

EXAMPLE 15—

16 semicircles = 8 circles = 8πd = 16π. So d = 2.

Each side of the rectangle (actually a square) is 8. 8 × 8 = 64, the area. The answer is B. Surprisingly, this part is now very easy. The area of the rectangle is <u>exactly</u> the area inside the regions, since for each semicircle inside, there is one outside. So the area is also 64. This is quite difficult to see in a short period of time, but that is one of the reasons I wrote this book:

to show you what to look for so that you can see the picture quickly, verry quickly.

EXAMPLE 16—

Revolution = 1 circumference = πd = 20π ft per revolution.

Total length is 500 ft.

$$\text{Revolutions} = \frac{500}{20\pi} = \frac{25}{\pi} \text{ revolutions.}$$

EXAMPLE 17—

This is not too bad. 2(160) = 320. 5(120) = 600. 600 + 320 = 920 feet of fencing.

Let's look at old Pythagorus and perhaps the most famous theorem in all of math!!!!

CHAPTER 15

OLD PYTHAGORAS (AND A LITTLE ISOSCELES)

The SAT has grown to ask many more questions about old Pythag. Most theorems have only 1 or 2 proofs. There are over 100 distinct proofs of this theorem. Here are the facts, just the facts.

Right triangle. Side c is the hypotenuse, opposite the right angle C. Sides a and b are called *legs.* In a right triangle, $c^2 = a^2 + b^2$. Of the original proofs, three were contributed by past presidents of the United States. We actually had presidents who knew and really cared about math!!!! Probably never again! A shame!!!!

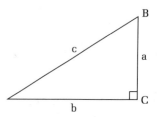

EXAMPLE 1—

$$x^2 = 5^2 + 7^2 = 74 \qquad x = \sqrt{74}$$

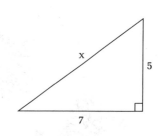

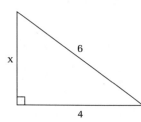

EXAMPLE 2—

$x^2 + 4^2 = 6^2$ (hypotenuse is allllways by itself.

$x^2 + 16 = 36$

$x^2 = 20.$

$x = \sqrt{20} = 2\sqrt{5}$

Isosceles triangle

Equal sides (2), AB = AC, are called *legs*. Angle BAB is a vertex angle. Base BC can be bigger than, equal to, or less than each leg. Base angle B = base angle C. AD, altitude, divides the base in half.

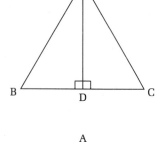

EXAMPLE 3—

Find the height x and the area of the triangle. (SAT likes this picture.)

AD wasn't originally given. If you draw it, BD = 4. If you notice, this is Example 2. $x = 2\sqrt{5}$

$A = \frac{1}{2}b \times h = \frac{1}{2} \times 8 \times 2\sqrt{5} = 8\sqrt{5}$ square units.

There are special right triangles the SAT loves. You must know them. I believe as the years go on they will show up more and more. As a bonus, for those of you who take geometry (most of you) and trig (hopefully more of you), they are a must!!

The exact Pythagorean triples (the most important ones)

1. The 3-4-5 group ($3^2 + 4^2 = 5^2$) 3, 4, 5   6, 8, 10
   9, 12, 15   12, 16, 20   15, 20, 25

2. The 5-12-13 group ($5^2 + 12^2 = 13^2$)
   5, 12, 13   10, 24, 26

3. The 8-15-17 group $(8^2 + 15^2 = 17^2)$ 8, 15, 17

4. The 7-24-25 group $(7^2 + 24^2 = 25^2)$ 7, 24, 25

If you are really bad at memorization, know at least
3, 4, 5,   6, 8, 10,   and   5, 12, 13.

There are two others that are vital. The first, the **45-45-
90 isosceles right triangle,** is one of the SAT absolute
favorites.

Legs are equal.

1. From leg to hypotenuse, multiply by $\sqrt{2}$.

2. From hypotenuse to leg, divide by $\sqrt{2}$.

EXAMPLE A—

Legs are equal. Soooo x = 7.

y, the hypotenuse $= x\sqrt{2} = 7\sqrt{2}$

That's it.

EXAMPLE B—

Hypotenuse to leg???? Divide by $\sqrt{2}$.

$$\text{Leg } x = y = \frac{8}{\sqrt{2}} = \frac{8\sqrt{2}}{\sqrt{2}\sqrt{2}} = \frac{8\sqrt{2}}{2} = 4\sqrt{2}$$

This is not too hard, but knowing this is vital. We need
to know one more triangle.

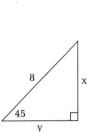

30°-60°-90° right triangle
Short leg is always opposite the 30° angle.
Long leg is allllways opposite the 60° angle.
Hypotenuse, the longest side, is always opposite the
90° angle.

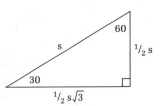

1. Always get the short leg first, if it is not given!!!

2. Short to hypotenuse, multiply by 2.

3. Hypotenuse to short, divide by 2.

4. Short to long, multiply by $\sqrt{3}$.

5. Long to short, divide by $\sqrt{3}$.

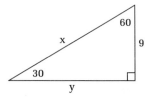

EXAMPLE C—

Short is given.

Short to hypotenuse, multiply by 2. $9(2) = 18$.

Short to long, multiply by $\sqrt{3}$. $9(\sqrt{3}) = 9\sqrt{3}$.

$x = 18$. $y = 9\sqrt{3}$.

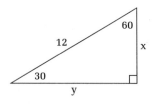

EXAMPLE D—

Hypotenuse is given. Find short first.

Hypotenuse to short, divide by 2. $12/2 = 6$.

Short to long, multiply by 3. $6\sqrt{3}$.

$x = 6$. $y = 6\sqrt{3}$.

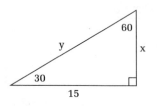

EXAMPLE E—

Long given. Get short first.

Long to short, divide by 3.

$$\frac{15}{\sqrt{3}} = \frac{15\sqrt{3}}{\sqrt{3}\sqrt{3}} = \frac{15\sqrt{3}}{3} = 5\sqrt{3}$$

Short to hypotenuse, multiply by 2.

$2(5\sqrt{3}) = 10\sqrt{3}$. $x = 5\sqrt{3}$. $y = 10\sqrt{3}$.

PROBLEMS INVOLVING THESE TRIANGLES ARE <u>VERY</u> EASY IF YOU KNOW THESE FACTS COLD. YOU MUSSSSTTTT KNOW THESE WELL!!!!

LET'S DO SOME PROBLEMS

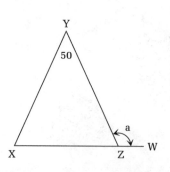

EXAMPLE 1—

$XY = YZ$. $a =$

EXAMPLE 2—

HJKM is a square. The area of this square is. . . .

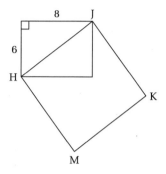

EXAMPLE 3—

A 25-ft ladder is placed against a vertical wall with the bottom of the ladder standing 7 feet from the base of the building. If the top of the ladder slips down 9 feet, then how far will the bottom slip?

A. 10 ft   B. 13 ft   C. 15 ft   D. 17 ft   E. 20 ft

EXAMPLE 4—

3 squares. WX is a line segment forming the diagonals of both smaller squares.

The ratio of $\dfrac{\text{length of WX}}{\text{length of YZ}} =$

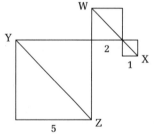

EXAMPLE 5—

Arc AB is ¼ of the circumference of the circle. Chord AB is y. What is the diameter of the circle?

LET'S SHOW SOME SOLUTIONS

EXAMPLE 1—

A relatively mild one to start. Angles X and XZY are equal, because it is an isosceles triangle. Call the angle a.

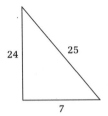

$2a + 50 = 180$ $2a = 130$ $a = 65$
a and b are sups $b = 180 - 65 = 115$

EXAMPLE 2—

If you know your Pythag trips, this is trivial. 6, 8, . . .
10!!!! $10^2 = 100$.

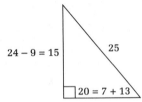
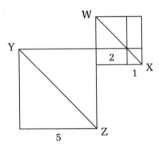

EXAMPLE 3—

If you know your trips <u>very well</u>, this is not too bad.

$x = 20$ (15, 20, 25 triple)

$20 - 7 = 13$ Answer is B.

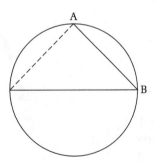

EXAMPLE 4—

A real trick. You do NOT need to know there are 45-45-90 triangles. Imagine that WX is the diagonal of a square that contains the smaller two squares. The ratio of the diagonals is the same as the ratio of the sides. The answer is just 3/5.

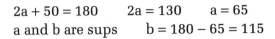

EXAMPLE 5—

A tricky, tricky problem. If you actually draw the arc in a circle, you would see it forms a 45-45-90 right triangle whose sides are y, y, and $y\sqrt{2}$, which is the diameter. $y\sqrt{2}$ is the answer.

You really have to know these triangles well, as you must know all the formulas, even though many are put at the top of the SAT. If you need to look, you are already in trouble.

LET'S DO SOME MORE!!!!

EXAMPLE 6—

BD is $\sqrt{8}$. The area of the 4 semicircles is. . . .

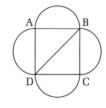

EXAMPLE 7—

A. x   B. 5

EXAMPLE 8—

PQ = QT = TR = RS. Angle y =

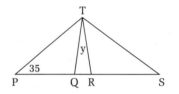

EXAMPLE 9—

AB = $6\sqrt{2}$. Area of shaded region is

EXAMPLE 10—

UVWX is a square. 2YW = VY. Angle a = ? degrees

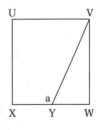

LET'S SEE THE RESULTS

EXAMPLE 6—

ABD is a 45-45-90 right triangle, hypotenuse BD. AB, the diameter of the semicircle = $\sqrt{8}/\sqrt{2} = \sqrt{4} = 2$.

Radius of semicircle is 1.

4 semicircles = 2 circles = $2 \times \pi(1)^2 = 2\pi$.

EXAMPLE 7—

If x = 5, $3^2 + 5^2 = 9 + 25 = 34$. Buuuut $6^2 = 36$, which is bigger. So x must be bigger.

EXAMPLE 8—

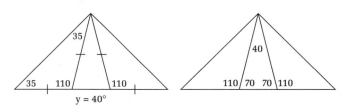

EXAMPLE 9—

The SAT really likes this picture, really really, really.

AO is the leg of a 45-45-90 right triangle and is $6\sqrt{2}/\sqrt{2} = 6$. The area of the shaded region is the area of ¼ circle minus the area of the triangle, where r = 6, b = 6, and, I know it's boring, but h = 6 also. Area of shaded region is

$$\tfrac{1}{4}\pi r^2 - \tfrac{1}{2}bh = \tfrac{1}{4}\pi(6)^2 = \tfrac{1}{2}(6)(6) = 9\pi - 18$$

EXAMPLE 10—

We have a 30-60-90 triangle. So angle WVY = 30°. So, angle VYW is 60°. a is the supplement = 120°.

I think we need some more. Remember geometry is about 20% of the SAT.

LET'S TRY SOME

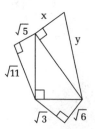

EXAMPLE 11—

$y^2 - x^2 =$

EXAMPLE 12—

XY ∥ OZ   Area of XYZO is

A. 12   B. 20   C. 30   D. 42
E. can't be determined

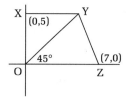

EXAMPLE 13—

A. x   B. y/4

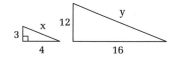

EXAMPLE 14—

a =

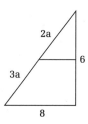

EXAMPLE 15—

A. length of hypotenuse of right triangle with legs 6
and 8   B. length of hypotenuse of right triangle with
legs 5 and 12

OK, LET'S SEE THE ANSWERS

EXAMPLE 11—

$z^2 = y^2 - x^2$.

But z^2 also $= a^2 + b^2$.

Buut $a^2 = (\sqrt{11})^2 + (\sqrt{5})^2 = 16$. $b^2 = (\sqrt{3})^2 + (\sqrt{6})^2 = 9$.

$a^2 + b^2 = y^2 - x^2 = 9 + 16 = 25$.

When you see it, it takes much longer to write out than
to do.

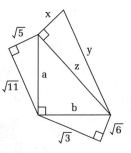

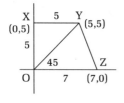

EXAMPLE 12—

45° angle   OX = 5.   So XY = 5 (also on a 45° line
x = y, which we'll look at a little later).

Area of trapezoid = ½h(b₁ + b₂)   h = 5   b₁ = 5
b₂ = 7   ½(5)(5 + 7) = 30   C.

The problems could be also done by finding the area of
the two triangles and adding, but this is better!!

EXAMPLE 13—

There are two ways to do this, both easy. The first, similar triangles.

3/12 = x/y   x/y = ¼

Soooo x = ¼y.   Answer is C.

Orrrr. 3, 4, 5 triangle, 12, 16, 20 right triangle, 5/20 =
1/4 = x/y and the answer still is C!

EXAMPLE 14—

Easy if you know the triple, 6, 8, . . . 10. 2a + 3a = 10.
a = 2.

EXAMPLE 15—

Absolutely trivial if you know the triples. 10 < 13.
Answer is B.

OK, A FEW MORE. . .

EXAMPLE 16—

CD is 6.   BD = 3.

Perimeter of the shaded region is

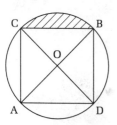

EXAMPLE 17

A right triangle has sides 1 and $\sqrt{2}$. What can the third side be?

I. 1   II. $\sqrt{2}$ III. $\sqrt{3}$

A. I only   B. II only   C. III only
D. I and II only   E. I and III only

EXAMPLE 18

Area of the semicircle is

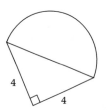

EXAMPLE 19

The ratio of NP to MN is

A. $\sqrt{2}-1$ B. $\sqrt{2}/2$ C. $\sqrt{2}$ D. $\sqrt{2}+1$ E. $2\sqrt{2}$

Last set (last set of this type).

LET'S LOOK AT THE ANSWERS

EXAMPLE 16

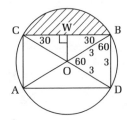

Tough problem! Lots of little things in the picture. It can be done quickly if you see the pieces. CD = 6 (diameter). CO = OB = radius = 3. So right most triangle is equilateral. All angles 60°. So angle OBW = angle WCO = 30°. So angle COW (moooo) = 60°. So triangle COW (mooo again) is a 30-60-90 triangle.

CO = 3   WO = 1.5   CW = 1.5$\sqrt{3}$
CB = 2CW = 3$\sqrt{3}$

Arc CB is 120°. 120/360 = ⅓ of circumference.
⅓(6π) = 2π. Total perimeter is 2π + 3$\sqrt{3}$.

EXAMPLE 17—

$$x = 1 \qquad 1^2 + 1^2 = (\sqrt{2})^2$$

Since 1 and 1 are usually 2, I is true.

$x = \sqrt{2}$ can't work no matter what order you try.

$x = \sqrt{3}$ works since $1^2 + (\sqrt{2})^2 = (\sqrt{3})^2$ since $1 + 2$ is usually 3. Answer is E.

EXAMPLE 18—

Hypotenuse = diameter of semicircle is $4\sqrt{2}$(45-45-90 right triangle).

$$r = 2\sqrt{2} \qquad A = \pi r^2 = \pi(2\sqrt{2})^2 = 8\pi$$

Area of half a circle is $\frac{1}{2}\pi r^2 = \frac{1}{2}8\pi = 4\pi$

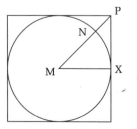

EXAMPLE 19—

This is truly a tough one, but the reason the problem is here is there is a sneaky way around it.

Let's do it the "right" way and the SAT way.

Let MX = 1. (You must know to draw MX.) XP = 1 (45-45-90 leg).

MP = $\sqrt{2}$. (45-45-90 hypotenuse.) MN is another radius = 1.

NP = MP − MN = $\sqrt{2}$ − 1. NP/MN = $\sqrt{2}$ − 1 / 1 = $\sqrt{2}$ − 1. Answer is A.

If you can do this, then you probably didn't need to buy this book. But suppose you can't do this. . . .

We can measure the ratio between NP and MN. Mark off NP on a pencil or scrap of paper. Then measure MN. You see that MN is a little more than twice NP. So NP/MN is a little less than ½. Let's look at the choices knowing that $\sqrt{2}$ = approx 1.4 ($\sqrt{3}$ = approx 1.7).

E. $2\sqrt{2} = 2(1.4)$ certainly is not less than ½.

D. $\sqrt{2} + 1 = 1.4 + 1$ surely is not less than ½.

C. $\sqrt{2} = 1.4$, nope.

B. $\sqrt{2}/2$ is about 0.7.

Possibly, we could have measured wrong, but $\sqrt{2} - 1 = 1.4 - 1 = .4$, yay, the correct answer. With choices, there sometimes is a trick like this.

I'm tired of these problems. Let's do something else. Pleeeeze!!!!!

CHAPTER 16

VOLUMES, SURFACE AREA, DISTANCE, DISTANCE BETWEEN POINTS

For the SAT, it is necessary to know the surface area and volume of a box and a cube. Actually, we need to know a little more, but we'll look at this in the section called "New Stuff" (relatively).

BOX

l = length; w = width; h = height

$V = l \times w \times h$

$SA = 2l \times w + 2w \times h + 2l \times h$

    bottom   sides   front
     + top   + back

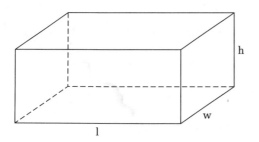

The only formula you might not want to remember is the surface area of a box. It is sometimes easier to look at the figure and add up the area of the rectangles you see.

EXAMPLE 1—

$V = 10 \times 5 \times 4 = 200$ cubic feet

$SA = 2(10 \times 4) + 2(5 \times 4) + 2(10 \times 5) = 220$ square feet

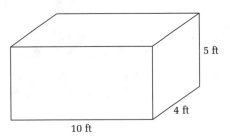

5 ft

4 ft

10 ft

CUBE

At certain times, the SAT loves the cube. You should know it well!!!!

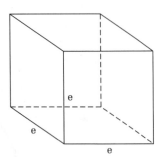

$e =$ edge.

$V = e \times e \times e = e^3$. (Cubing comes from a cube?!! Logical, isn't it?!)

$SA = 6e^2$.

There are 12 edges.

There are 8 vertices.

There are 6 faces (surfaces).

Diagonal in a face. Length $= e\sqrt{2}$.

Diagonal of the cube (one corner on top to opposite corner on bottom). $e\sqrt{3}$.

3-D Pythagorean theorem:

$a^2 + b^2 + c^2 = d^2 \qquad d^2 = e^2 + e^2 + e^2 = 3e^2$

Sooo $d = e\sqrt{3}$.

EXAMPLE 2

$V = 10^3 = 1000$ cubic feet

$SA = 6(10^2) = 600$ square feet

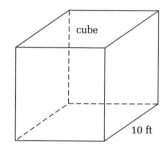

cube

10 ft

POINTS ON A GRAPH

Now let's talk about points on a graph. Normally,

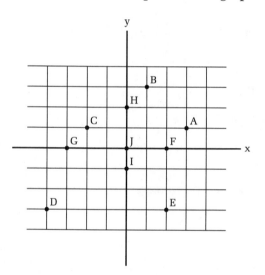

graphing points is about as easy a topic as there is. But the SAT sometimes makes the questions kind of tricky.

First, you must know how to locate points.

A(3,1) B(1,3) Notice order makes a difference.

C(−2,1) D(−4,−3) E(2,−3).

G(−3,0) F(2,0) For any point on x axis, y coordinate is 0!

I(0,−1) H(0,2) For any point on y axis, x coordinate is 0!

J(0,0) Origin.

If the SAT stuck to questions like this, most would be very easy. But. . . .

The graph is divided into four regions called *quadrants.*

In quadrant I, both x and y are positive. See points A and B in the graph on p. 119.

In quadrant II, y is positive but x is negative. See point C.

In quadrant III, both x and y are negative. See point D.

In quadrant IV, x is positive but y is negative. See point E.

F, G, H, I, J are on the axes and are not in a quadrant.

There is a little more you must know.

Relative Size of x and y

This is something the SAT asks.

In the shaded area, the y coordinate is bigger than the x coordinate.

In the nonshaded area, the x coordinate is bigger than the y coordinate.

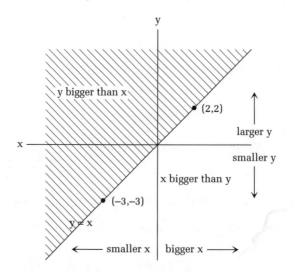

x is sometimes called the *first coordinate.* Guess what y is called.

On the line y = x, the x and y values are equal.

The higher the point, the higher the y value.

The righter the x, the higher the x value.

You may think all of this is silly, but sometimes it gets tricky. Here's a little more.

Horizontal lines are y = something.

x axis, a horizontal line, is y = 0.

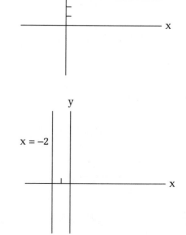

Vertical lines are x = something.

y axis, a vertical line, is x = 0.

One-Dimensional Distances

Because the y values are the same, the length of the line segment is the right x minus the left x, 7 − 2 = 5.

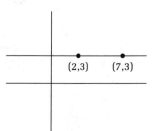

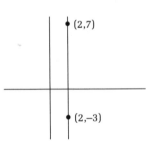

Because the x values are the same, the length of the line segment is top y value minus bottom y value, $7 - (-3) = 10$.

Two-Dimensional Distance Formula— Old Pythag Again

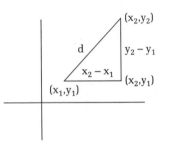

$$d = \sqrt{(x_2 - x_1)^2 + (y_2 - y_1)^2}$$

Find the distance between $(4,-2)$ and $(9,5)$

$$\begin{array}{cccc} x_1 & y_1 & x_2 & y_2 \end{array}$$

$$d = \sqrt{(9 - 4)^2 + (5 - (-2))^2} = \sqrt{74}$$

Symmetry

If the point (x,y) is in I, then

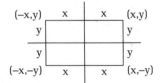

$(-x,y)$ is in II

$(-x,-y)$ is in III

$(x,-y)$ is in IV

LET'S FINALLY TRY SOME PROBLEMS

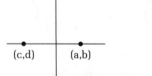

EXAMPLE 1 (2 PARTS)

A. c   B. a

A. d   B. b

EXAMPLE 2

How many cubes, 2″ on each side, can be put in this box?

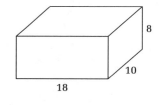

8

10

18

EXAMPLE 3

ABCD is a square.

A. area of square   B. 10

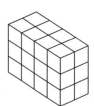

C (3,4)

(0,3) B

C (3,4)

D (4,1)

A (1,0)

EXAMPLE 4

The box is divided into cubes 10 inches on each edge. Find the volume of the box.

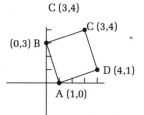

EXAMPLE 5

Box G: volume 20. Box H: volume 40.

A. area of base of G   B. area of base of H

LET'S LOOK AT THESE ANSWERS

EXAMPLE 1

Part 1: On the x axis, a > c since a positive is bigger than a negative.   B.

Part 2: On the x axis, all y values are = (=0).   C.

EXAMPLE 2

The volume of a $2 \times 2 \times 2$ cube is 8. Annnnd $18 \times 10 \times 8$ divided by 8 (8s cancel) is $10 \times 8 = 180$.

Orrr $\dfrac{18}{2} \times \dfrac{10}{2} \times \dfrac{8}{2} = 9 \times 5 \times 4 = 180$.

EXAMPLE 3

Because we are told it's a square, all we have to do is find one side and square it. Even better, let's just use the square of the distance formula.

Take the easiest two points: (0,3) and (1,0)

$(x_2 - x_1)^2 + (y_2 - y_1)^2 = (0 - 3)^2 + (1 - 0)^2 = 9 + 1 = 10$. C, equal.

EXAMPLE 4

This is almost Example 2. 2 cubes by 3 cubes by 4 cubes means . . . the dimensions are $20 \times 30 \times 40 = 24,000$ cubic units.

EXAMPLE 5

This is a question where, if you look really quickly, you will miss it. The answer is D. The base could be anything in either, for example . . .

Box G: The dimensions could be
$1 \times 1 \times 20$.   Area of the base is 1.   orrrr

$5 \times 4 \times 1$.   Area of the base is 20.

Box H: The dimensions could be
$1 \times 1 \times 30$.   Area of the base is 1.   orrrr

$5 \times 2 \times 3$.   Area of the base is 10.   orrrr

$6 \times 5 \times 1$.   Area of base is 30.

NEXXXXT. . . .

EXAMPLE 6

A circle is tangent (just touching) the y axis and has center $(-5,-7)$. Its radius is

A. −5   B. 5   C. −7   D. 7   E. $\sqrt{74}$

EXAMPLE 7—

L is the line y = x.

A. c   B. d

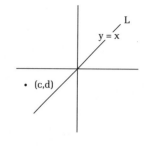

EXAMPLE 8—

Surface area of the box is

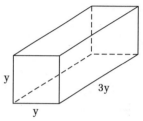

EXAMPLE 9—

A cube has edge 4. Find the length of the line segment from the middle of one face to a vertex of the opposite face.

EXAMPLE 10—

The length of MN is

LET'S LOOK AT HOW TO DO THESE

EXAMPLE 6—

You must draw a picture and see the radius is 5.
B. Lengths can't be negative.

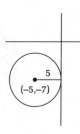

EXAMPLE 7—

In the region of (c,d), we know the y value is bigger.   B.

EXAMPLE 8—

This is a straight surface area problem. Two squares at the end. . . . area is $2y^2$. Other 4 sides, $3y^2$ . . . area is $12y^2$, total $14y^2$.

EXAMPLE 9—

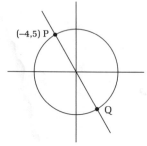

Pretty tough one. If the edge is 4, AD is also 4. CD is ½ of 4 = 2.

$AC^2 = AD^2 + CD^2$ $AC^2 = 2^2 + 4^2 = 20$ BC also = 2.

$AB^2 = BC^2 + AC^2 = 4 + 20$ $AB = \sqrt{24}$

EXAMPLE 10—

Straight distance formula.

$$\sqrt{(5-1)^2 + (4-1)^2} = \sqrt{25} = 5.$$

LET'S TRY FIVE MORE

EXAMPLE 11—

The lengths of all sides of a rectangular solid are integers. If the volume is 17, find the area of all of its surfaces.

EXAMPLE 12—

The coordinates of Q are. . . .

EXAMPLE 13—

A cube has a volume of 72 cubic zrikniks. It is divided into 8 ≅ cubes. Find the ratio of the edge of the new cube to the edge of the old cube.

EXAMPLE 14—

If the distance from (3,x) to (7,0) is 5, x may be

A. 0   B. 1   C. 2   D. 3   E. 4

EXAMPLE 15—

Area of triangle OC is. . . .

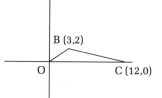

LET'S ANSWER THESE FIVE

EXAMPLE 11—

It doesn't say distinct integers. The only possibility is 1 by 1 by 17.

Two ends are 1 by 1 squares. The total area is 2.

Four others are 1 by 17 rectangles. $17 \times 4 = 68$; $68 + 2 = 70$.

EXAMPLE 12—

By symmetry, Q must be (4,–5).

EXAMPLE 13—

We don't care about the measurements or the planet it is located in. The ratio of

$v_1/v_2 = e_1^3/e_2^3 = 1/8$

So $e_1/e_2 = 1/2$.

Notice, we don't need to know what either edge is, just the ratio.

EXAMPLE 14—

$(7 - 3)^2 + (0 - x)^2 = 25$. $x^2 + 16 = 25$. $x^2 = 9$. x could be 3.   D.

Trial and error will work also.

EXAMPLE 15—

Base of triangle is 12. Height of triangle is 2, the y value of (3,2). (You should see it.)

$A = \frac{1}{2} \times 2 \times 12 = 12$.

NEXT FIVE ARE. . . .

EXAMPLE 16—

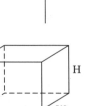

Point Q has coordinates (a,b). If m > a > b > n, which points could have coordinates (m,n)?

A. V   B. W   C. X   D. Y   E. Z

EXAMPLE 17—

L, W, H are dimensions of a rectangular box and all measurements are distinct integers greater than 1. The volume could be

A. 8   B. 12   C. 24   D. 27   E. 98

EXAMPLE 18—

Quadrants satisfying $x/y = 3$.

A. I only   B. I and II   C. I and III   D. II and IV
E. All of them

EXAMPLE 19—

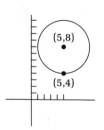

Circumference of this circle is

EXAMPLE 20—

A. b/a   B. d/c

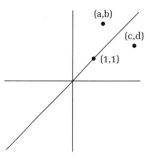

LET'S LOOK AT THESE ANSWERS

Remember, geometry is a pretty big part of the SAT.

EXAMPLE 16—

m > a could be X, Y, or Z, a bigger x value.

b > n could only be X, a smaller y value.   C.

EXAMPLE 17—

The only possibility of 3 distinct factors (not necessarily primes because it doesn't say primes) is 24 = $2 \times 3 \times 4$.   C.

EXAMPLE 18—

The number 3 is misleading. x/y = positive, which means either x and y are both positive or x and y are both negative. I and III.   C.

EXAMPLE 19—

Radius is $8 - 4 = 4$. $c = 2\pi r = 2\pi(4) = 8\pi$.

EXAMPLE 20—

All numbers are positive.

In the upper left, y is bigger than x. y/x > 1.

In the lower right, x is bigger than y. d/c < 1.

Sooooo y/x is bigger. The answer is A.

LET'S DO A FINAL FIVE

These are the last five in this section.

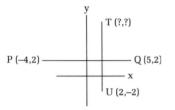

EXAMPLE 21

$PQ = TU$ (a little rhyme) and PQ is perpendicular to TU. The point T is

A. (2,6)   B. (2,7)   C. (2,8)   D. (7,2)   E. (8,2)

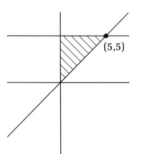

EXAMPLE 22—

Which point is in the shaded area?

A. (2,4)   B. (−2,4)   C. (2,6)   D. (4,2)   E. (6,2)

EXAMPLE 23—

Find the volume of a cube if the surface area is $54a^2$ square units.

A. $9a^2$ B. $27a^3$ C. $81a^2$ D. $81a^3$ E. $729a^3$

EXAMPLE 24—

Find the ratio of the volume of a cube with edge $\sqrt{2}$ to the volume of a cube with edge $\sqrt[3]{2}$.

EXAMPLE 25—

The coordinates of Q are

A. (a,−b)   B. (b,a)   C. (b,−a)   D. (−a,b)
E. (−a,−b)

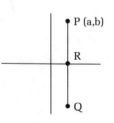

THE ANSWERS TO THE LAST FIVE ARE. . . .

EXAMPLE 21—

PQ. The y values are the same. The length of the line is the difference in the x's. $5 - (-4) = 9$. TU = PQ and perpendicular. The perpendicular parts mean the x numbers are the same $(2,?)$. The equal part means the length of the lines are the same. The y number of T is 9 more than -2, which is 7. Answer is. . . . BBBB.

EXAMPLE 22—

The shaded area has x and y < 5 annnnd y is bigger than x annnnd both coordinates are nonnegative. A is the answer.

EXAMPLE 23—

$$SA = 6x^2 = 54a^2$$

$$x^2 = 9a^2$$

$$x = 3a$$

$$V = x^3 = (3a)^3 = 27a^3 \qquad B.$$

EXAMPLE 24—

Ratio of volumes is $(\sqrt{2})^3/(\sqrt[3]{2})^3 = 2\sqrt{2}/2 = \sqrt{2}$.

EXAMPLE 25—

Vertical lines have the same x values. y is negative. $(a,-b)$. Answer is A.

As much as I would like more, I think it's time to go on to ratios.

CHAPTER 17

THE RATIO IS . . .

Hi. I've just taken a mini-, minibreak from the geometry section. The section on ratios overlaps many other topics from the past, as do a number of others. We have already talked about ratio and proportion, which is two ratios equal to each other. What really is a ratio? It is simply a comparison of two items. How do we represent ratios? There are two ways: a new fashion (relatively) and old fashion.

EXAMPLE 1—

Write the ratio of 2 to 5.

"New" way: as a fraction, 2/5

Old way: with a colon, 2:5

EXAMPLE 2—

Find the ratio of 5 inches to 2 feet. Answer, 5/24.

The measurements must be the same and 12 inches = 1 foot, as I hope you know.

EXAMPLE 3

If 20 items cost 46¢, how much do 30 items cost?

$$\frac{\text{cost}}{\text{item}} = \frac{\text{cost}}{\text{item}} \qquad \frac{46}{20} = \frac{x}{30}$$

Cross multiply

$20x = 46(30)$. Soooo $x = 46(30/20) = 69$.

Even with a calculator, this takes too much time. So we can use a trick. Both 20 and 30 are multiples of 10.

20 cost 46¢

10 cost 23¢ (divide by 2)

30 cost 69¢ (multiply by 3)

I love this stuff. By the way, hopefully you can do this in your head. Remember, you don't want to write anything if possible.

With some problems, you don't even need to know what you are talking about. Let's do one.

PROBLEMS

EXAMPLE 4

If 5 brigs equal 7 grigs, how many brigs = 11 grigs?

$$\frac{\text{brigs}}{\text{grigs}} = \frac{\text{brigs}}{\text{grigs}} \qquad \frac{5}{7} = \frac{x}{11} \qquad x = 5(11)/7 = 55/7.$$

Simple. Huh?

LET'S TRY SOME REAL (REAL???) PROBLEMS

EXAMPLE 1

The cost of 700 items at $1.50 per 100.

EXAMPLE 2—

A wire of uniform density and composition weighs 32 pounds. It is cut into 2 pieces. One is 60 feet and 24 pounds. What is the length of the original piece?

EXAMPLE 3—

Of 90 students in a class, the ratio of boys to girls is 2:3. How many girls?

EXAMPLE 4—

Five oranges cost c cents. How many can be bought for y dollars?

EXAMPLE 5—

If x is k% of y, what % of y is kx?

EXAMPLE 6—

In a certain garden, 1/5 of the flowering plants represent 1/10 of all the flowering plants. What is the ratio of flowering plants to nonflowering plants?

EXAMPLE 7—

A jar contains blue and red marbles, 20 in all. Each of the following can be the ratio of blue to red <u>except</u>

A. 1:1   B. 3:2   C. 4:1   D. 5:1   E. 9:1

LET'S SOLVE THESE

EXAMPLE 1—

A simple ratio problem that can be done without ratio.

700/100 = 7 times the amount. 7 × $1.50 = $10.50.

EXAMPLE 2—

Ratio trick, trick, trick, trick!!!! Both 32 and 24 are multiples of 8.

60 ft = 24 pounds

20 ft = 8 pounds (divide by 3)

80 ft = 32 pounds (multiply by 4 orrrr add the two together)

EXAMPLE 3—

Easiest way is 2x + 3x = 90. 2x = boys and 3x = girls. 5x = 90. x = 18. 3x = 3(18) = 54 girls.

EXAMPLE 4—

$$\frac{cost}{orange} = \frac{cost}{orange} \qquad y \text{ dollars} = 100y \text{ cents}$$

$$\frac{c}{5} = \frac{100y}{?} \qquad ? = 500y/c$$

EXAMPLE 5—

$$x = \frac{k}{100}\,y \qquad x = \frac{ky}{100} \qquad kx = k\,\frac{ky}{100} = \frac{k^2}{100}\,y.$$

k^2 percent.

EXAMPLE 6—

1/5 flowering   1/10 total.

1 flowering   5/10 total (½ total).

½ flowering and ½ not flowering. Ratio is 1/1!!!! Careful!!!!

EXAMPLE 7—

1:1 OK (10 and 10); 3:2 OK (12 and 8); 4:1 possible (16 and 4); 9:1 OK (18 to 2). But D is not possible because 5:1 (5 + 1 = 6) is no good since 20 is not a multiple of 6.

LET'S TRY SEVEN MORE

EXAMPLE 8—

If 2/3 of the perimeter of an equilateral triangle is 12, the perimeter is . . .

A. 8   B. 16   C. 18   D. 24   E. 36

EXAMPLE 9—

On a map, two cities that are 2.4″ apart are really 12 miles apart. What is the length of a .2″ straight road?

EXAMPLE 10—

Sugar costs m cents per pound. How many pounds can be bought for 6 dollars?

EXAMPLE 11—

In a 10-pound solution of water and alcohol, the ratio by mass of water to alcohol is 3:2. A 6-pound solution consisting of 2 parts water to one part alcohol is added to the 10-pound solution. What fraction of the new solution is alcohol?

EXAMPLE 12—

Q and R are two points to the right of A on the number line.

$2AQ = 3AR$. What is RQ/AR?

A

EXAMPLE 13—

$a^4 = 5$. $a^3 = 2/c$. Write a in terms of c.

EXAMPLE 14—

At a certain school, a liters of milk are needed per week per student. At that rate, b liters will supply c students for how many weeks?

LET'S LOOK AT THE ANSWERS

EXAMPLE 8—

Only an SAT would ask a question like this. But it is not hard. It doesn't matter that it's an equilateral triangle. $2/3\ p = 12$. $1/3\ p = 6$. $p = 18$. Hopefully done in your head.

EXAMPLE 9—

$.2/.24 = 1/12$ of $12 = 1$.

EXAMPLE 10—

Same as an earlier one.

$$\frac{m}{1} = \frac{600 \text{ (cents)}}{?} \qquad ? = 600/m \text{ pounds}$$

EXAMPLE 11—

10 lb 3:2 water   6 lb water   4 lb alcohol

6 lb 2:1 water $\quad \dfrac{4 \text{ lb water}}{10 \text{ lb water}} \quad \dfrac{2 \text{ lb alcohol}}{6 \text{ lb alcohol}}$ 16 lb total

$6/16 = 3/8$ alcohol

A   R   Q

Reason: $2 \times 3 = 3 \times 2$

EXAMPLE 12—

$2AQ = 3AR$. Let Q be 3 units from A and R 2 units from A as shown in the following figure. Clearly, the figure shows $RQ/AR = 1/2$.

EXAMPLE 13—

$$\frac{a^4}{a^3} = \frac{5}{2/c}$$

Neat trick. $a = 5c/2$.

EXAMPLE 14—

I think this is a toughie. We'll work it with both numbers and letters. Suppose a = 6 liters for one student for
one week. Two students for one week would be 12. Or,
one student for three weeks would be 18 liters. So, if
you multiply or divide the students or weeks, leaving
the other alone, you would mult (divide) the liters by
the same number. Let c = students = 7. b = 84.

Liters	Students	Weeks	Liters	Students	Weeks
6	1	1	a	1	1
42	7	1	ac	c	1
1	7	1/42	ac/ac = 1	c	1/ac
84	7	(1/42)84	1(b) = b	c	b(1/ac) = b/ac
					is the answer

This is a particularly tough one, but most ratio problems aren't.

LET'S DO A FINAL SEVEN

EXAMPLE 15—

A gas tank on a tractor holds 18 gallons. A tractor
needs 7 gallons to plow three acres. How many acres
does the tractor plow with a full gas tank?

EXAMPLE 16—

A penguin swims at 8 meters per second. How long
does it take the penguin to swim 100 meters?

EXAMPLE 17—

If z is 70% of y, and x is 60% of y, the ratio of z to x is

EXAMPLE 18—

It takes 10 people 12 hours to do a job. How many
hours will it take for 6 people to do ¼ of the job?

EXAMPLE 19—

$$\frac{b+a}{b} = 2 \qquad \frac{c+a}{a} = 3 \qquad b/c =$$

EXAMPLE 20—

The ratio of Sandy's salary to Chris's salary is 2 to 5. The ratio of Sandy's salary to Cory's salary is 7 to 9. Find the ratio of Chris's salary to Cory's.

EXAMPLE 21—

What is the thickness of one sheet of paper if 500 sheets is 2.5 inches thick?

LET'S LOOK AT THE LAST SEVEN ANSWERS

EXAMPLE 15—

Straight proportion. 7/3 = 18/a. Acres a = 54/7 acres.

EXAMPLE 16—

SAT distance problem up to this point. rate × time = distance. Soooo, distance over rate = time. 100/8 = 12.5 seconds.

EXAMPLE 17—

$z = .7y$

$x = .6y$

Divide!!!! $\dfrac{z}{x} = \dfrac{.7}{.6} = 7/6$ That's it.

EXAMPLE 18—

You have to look at it as 120 hours to finish one job; 30 hours for ¼ of the job. Six people would take 5 hours to finish ¼ of the job. (6)(5) = 30.

EXAMPLE 19—

$$\frac{b+a}{b} = \frac{b}{b} + \frac{a}{b} = 2 \qquad \text{So } \frac{a}{b} = 1$$

$$\frac{c+a}{a} = \frac{c}{a} + \frac{a}{a} = 3 \qquad \text{So } \frac{c}{a} = 2$$

$$\frac{a}{b} \times \frac{c}{a} = \frac{c}{b} = 2 \times 1 = 2 \qquad \frac{b}{c} = 1/2$$

EXAMPLE 20—

Similar problem.

$$\frac{S}{Ch} = \frac{2}{5} \qquad \frac{S}{Co} = \frac{7}{9}$$

$$\frac{Ch}{Co} = \frac{Ch}{S} \frac{S}{Co} = \frac{5}{2} \times \frac{7}{9} = \frac{35}{18}$$

There are other ways, such as problem 17.

EXAMPLE 21—

Simple ratio, but the larger number is on the bottom.

$$\frac{2.5}{500} = \frac{5}{1000} = .005$$

Let us take a look at the trends of the SAT since 1994. We have already mentioned the increase in algebra, especially exponents. So far I think there are four trends: Two I think are kind of easy; a third I think will rarely come up, and a fourth. Let's look, comparing them to what has come before.

CHAPTER 18

FROM OLD STUFF TO NEW (RELATIVELY)

CYLINDERS

The SAT has added the formula of the volume of a cylinder to the list of formulas you need to know. I have seen more than 60 past SAT and PSAT exams. On none of them did you need to know the formula.

I have a crazy head. I can make up SATs. But to this point I find it very difficult to make up questions about the volume of the cylinder. The SAT has had a few, but they are indeed strange. I suspect these kinds of questions will be few and far between. Of course, at most one will appear on any SAT. I don't think you should worry about it. Let me give you the formula, a straightforward question, and an SAT-type one.

Volume = $\pi r^2 h$
r = radius of the circular base and h = height

PROBLEMS

EXAMPLE 1—

Find the volume of a cylinder if the diameter of the base and the height are both 6.

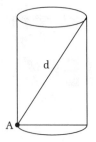

EXAMPLE 2—

The radius of the base of a cylinder is 4. From a point furthest from point A on the base is another point. If the height is 6, find the distance to that point.

SOLUTIONS

EXAMPLE 1—

$d = 6$ $r = 3$ $V = \pi r^2 h = V = \pi(3)^2 6 = 54\pi$

EXAMPLE 2—

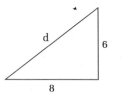

If you look at the picture, we have a simple Pythag triple, 6, 8, . . . 10!!

SLOPE

The slope of a line has been indirectly talked about before. It seems to be more in the consciousness of those who make up the SAT. Let us define slope formally.

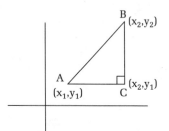

$$\text{slope } m = \frac{y_2 - y_1}{x_2 - x_1} = \frac{\text{change in } y}{\text{change in } x}$$

EXAMPLE 1—

   a. Find the slope between (2,3) and (5,7).

   b. Find the slope between (3,–4) and (–1,1).

   c. Find the slope between (1,5) and (4,5).

   d. Find the slope between (2,2) and (2,6).

If you walk from left to right and you get to the line and you go up, the slope is POSITIVE, like 1(a).

If you go left to right and the line is at your head, this is a NEGATIVE slope, like 1(b).

1b. $m = \dfrac{1 - (-4)}{-1 - 3} = \dfrac{5}{-4}$

$= -\dfrac{5}{4} = \dfrac{-5}{4}$

1a. $m = \dfrac{7 - 3}{5 - 2} = \dfrac{4}{3}$

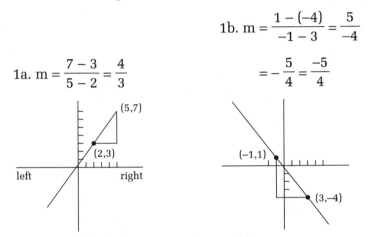

Horizontal lines have ZERO slope 1(c).

Vertical lines have NO SLOPE or the slope is said to be UNDEFINED 1(d).

1c. $\dfrac{5 - 5}{4 - 1} = 0/3 = 0$

1d. $\dfrac{6 - 2}{2 - 2} = \dfrac{4}{0}$ undefined

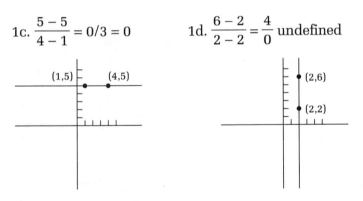

One form of the equation of a line is $y = mx + b$. The coefficient of x, m, is the slope. The y intercept is $(0,b)$. If $x = 0$, $y = b$.

If $y = -2x + 4$, the slope is -2 and the y intercept is 4, the point $(0,4)$.

If $y = x$, the slope is 1 (since $x = 1x$) and the y intercept is the origin $(0,0)$.

LET'S DO SOME PROBLEMS......

EXAMPLE 1—

A line is connecting (0,0) and (12,16). Another point on this line is

A. (2,3)   B. (3,2)   C. (4,3)   D. (3,4)   E. (16,12)

EXAMPLE 2—

If the points (1,2) and (x,8) are on a line with the slope 2, x might be

A. 2   B. 3   C. 4   D. 5   E. 6

EXAMPLE 3—

The slope of the line tangent (just touching) the circle at the point (0,2) is

A. 0   B. 2   C. undefined   D. 1
E. unable to determine the slope

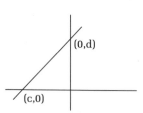

EXAMPLE 4—

The slope of this line is

A. c/d   B. −c/d   C. d/c   D. −d/c   E. cd

(0,d)

(c,0)

SOLUTIONS

EXAMPLE 1—

m between points given is

$$\frac{16 - 0}{12 - 0} = 16/12 = 4/3$$

D also has the same slope with (0,0).

EXAMPLE 2—

$$\frac{8-2}{x-1} = 2/1 \qquad 6 = 2(x-1) \qquad 2x = 8. \qquad x = 4 \qquad C.$$

EXAMPLE 3—

Horizontal line. m = 0.

EXAMPLE 4—

$$m = \frac{d-0}{0-c} = d/(-c) = D$$

Let's go on.

GRAPHS AND CHARTS

The SAT still has its share of charts and graphs. There are the usual ones and the strange ones. To these the SAT seems to have added ones I've seen on the GRE exam (Graduate Record Exam, the one that college seniors take to get into graduate school). These charts concern percentages on charts. We'll do one here and at least one on one of the practice SATs.

Problems

EXAMPLE 1—

We take the sum of numbers x + y, where x is taken from column X and y is taken from column Y. The total number of distinct sums is

A. 4   B. 6   C. 7   D. 8   E. 16

X	Y
2	4
3	5
4	6
5	7

EXAMPLE 2—

$2100 is Sandy's monthly budget. Of the $2100, what is spent on rent?

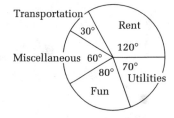

EXAMPLE 3—

A. % inc from 09 to 10   B. % inc from 09 to 10

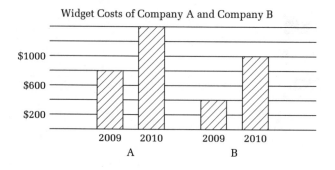

Widget Costs of Company A and Company B

LET'S LOOK AT SOME ANSWERS

EXAMPLE 1—

It does not matter what makes up the sum. 3 + 5 and 4 + 4 give the same result, 8. There are 7 distinct sums 6, 7, 8, 9, 10, 11, 12. Answer is C.

EXAMPLE 2—

$120° = 120/360 = 1/3$ of the circle. 1/3 of $2100 is $700.

EXAMPLE 3—

This is a really mild example of percent differences.

Company A: The increase is from $800 to $1400, an increase of $600.

Company B: The increase is from $400 to $1000, an increase of $600.

Both increases are the same, but C is not the answer. We are interested in % increases. % increases are based on the original amount (also % decreases).

Company A: $600/$800 = 75%.

Company B: 600/400 = 150%.

Answer is B.

NOTE

It is much better to work with fractional increases than % increases, even if the problem says so. Of course, with some problems you can't.

$$\frac{600}{800}, \frac{600}{400}$$

Tops the same, the bigger the bottom, the smaller the fraction.

So $\frac{600}{400}$ is bigger.

"STORY" WORD PROBLEMS

I have seen about 60 SATs, maybe a few more. Until about 1994, I can remember only one regular "story" word problem. After the change in format, I think I've seen four or five. Therefore, we'll look at the old-time word problems, which are still present, and the new ones. But you should NOT worry about them. I have never seen an SAT with more than one of these. If you get 1 wrong and 59 right, you'll get a pretty good score.

Age Problems

The older kind: Mary was d years old b years ago. How old will she be in c years?

The newer kind: John is 4 times the age of Mary. In 5 years he will be 3 times as old as Mary. What are the ages now?

The older kind solution: The secret is the age now. b years ago, Mary's age was d. Now, b years later, her age is d + b. c years in the future? d + b + c. If it were m years ago you were looking for, it would have been d + b − m.

You could also substitute numbers, but this is much quicker.

The newer kind solution: A chart will be helpful. On the SAT don't use lines—takes too much time. Let x = smaller number, Mary's age.

	Age now	Age in 5 years
John	4x	4x + 5
Mary	x	x + 5

In 5 years he   will be   3 times as old as Mary
4x + 5   =   3(x + 5)

$4x + 5 = 3x + 15$. $x = 10$ Mary's age. $5x = 50$ John's age That's it.

Coin Problems (Also Ticket Problems)

Old style: Find the total amount of money if you have n nickels, d dimes, and q quarters.

New style: There are 20 coins in dimes and nickels. The total is $1.70. How many dimes?

Both problems depend on the same things. 7 nickels is $7(5) = 35$¢. 8 dimes is $8(10) = 80$¢ in value. In other words, value of a coin (or ticket) times the number of coins (or tickets) is total $.

Old style: Value of n nickels is 5n; d dimes? 10d; q quarters? 25q. Total value is $5n + 10d + 25q$.

New style:

	Value of a Coin	× Number of Coins	= Total Amount of Money
nickels	5	20 − x	5(20 − x)
dimes	10	x	10x
totals		20	170

$5(20 - x) + 10x = 170$

$100 - 5x + 10x = 170$

$5x = 70$ $x = 14$ dimes.

Never check on the SAT. Not enough time.

NOTE I

Total money is in pennies.

NOTE 2

20 coins. If x is one part, 20 − x in the other. (If one were 6, the other would be 20 − 6.)

Alcohol Problems

Alcohol problems are the same. We've done the older kind. Oh, well let's do one.

New: How many ounces of 40% alcohol must be mixed with 6 oz of 70% alcohol to give a 50% solution?

If something is 40% alcohol and we have 20 lb, $.40 \times 20 = 8$ lb of alcohol. Okay, alcohol doesn't come in pounds, but who cares? You get the idea.

amount of a solution x% alcohol = amount of alcohol

Since all three (A, B, and Mixture) have %, we can eliminate the decimal point and the problem still works.

	Amount of Solution ×	Percent Alcohol =	Amount of Alcohol
A	x	40	40x
B	6	70	420
Mixture (total)	x + 6	50	50(x + 6)

$$40x + 420 = 50(x + 6)$$
$$40x + 420 = 50x + 300$$
$$120 = 10x$$
$$x = 12$$

Pleeeezzzze, do NOT worry too much about these. You should work on getting all the ones you know right. You'll get a good score that way.

CHAPTER 19

MISCELLANEOUS

This section is the reason it is impossible to teach any-one to get 800. Basically, this section would have to be 500 pages for you to have a chance for 800. But the SAT still asks questions that were never asked before and never will be again. If you can figure them out, and get everything else correct, you might be able to get 800.

PROBLEMS

EXAMPLE 1

There are eight people in a circle. Each shakes each other's hand once. How many handshakes?

EXAMPLE 2

A lunch consists of 1 sandwich, 1 soup, and 1 drink. The choice is from 6 soups, 3 sandwiches, and 2 drinks. How many different meals are possible?

EXAMPLE 3

A bowl has 7 red balls and 5 yellow balls.

a. One ball is selected. Find the probability the ball is yellow.

b. Two balls are selected without replacement. Find the probability that both balls are yellow.

c. Two balls are selected with replacement. Find the probability that both balls are red.

SOLUTIONS

EXAMPLE 1—

There are two ways to look at this.

METHOD 1
Person 1 shakes 7 peoples' hands. Person 2 shakes 6 peoples' hands, because 1 person's hand was shook already.

$7 + 6 + 5 + 4 + 3 + 2 + 1 = 28$

METHOD 2
Each person shakes 7 hands. $8 \times 7 = 56$. Buuut each pair is counted twice (person 3 shakes 4 and 4 shakes 3). ½ of $56 = 28$.

EXAMPLE 2—

The answer is $6 \times 3 \times 2 = 36$. This is called the *principal of counting*. If you can do a_1 in m ways, a_2 in n ways, a_3 in p ways, . . . The total number of ways you can do a_1 first, then a_2 second, a_3 next . . . is $m \times n \times p \times$ . . .

EXAMPLE 3—

Probability = total successes over total number. y =

a. Pr (yellow) = 5/12

b. p(y) = 5/12. No replacement pr (2nd yellow) = 4/11. (1 yellow gone; one ball gone.) p (1st yellow, then 2nd yellow) = (5/12)(4/11) = 5/33.

c. Watch. We changed to red. Pr (red, then red, replacement) = (7/12)(7/12) = 49/144.

Notice in b and c we are using Example 2.

LET'S DO SOME MORE

EXAMPLE 4—

We have the sequence 1, 2, 3, 4, 5, 1, 2, 3, 4, 5, 1. . . . The 798th entry is

    A. 1   B. 2   C. 3   D. 4   E. 5

EXAMPLE 5—

Suppose $a * b = b^2 - 5a$

  I. What is 2 * 10?

  II. If $b * b = 6$, what are the values of b?

SOLUTIONS

EXAMPLE 4—

This is a different kind of counting problem. Sometimes in mathematics, counting is easy, sometimes it is hard, and sometimes it is tricky. This sequence is cyclic (goes in a cycle). The cycle is 5. Soooo $5\overline{)798}$. We don't care about the answer. The remainder is 3. Answer is C.

EXAMPLE 5—

The SAT used to love, love, love this kind of problem. Now it only loves them.

It is a made up operation. Your job is to follow directions.

  I. $a * b = b^2 - 5a$. These directions tell you to square the number after the * and subtract 5 times the number before the * sign.

    $2 * 10 = 10^2 - 5(2) = 100 - 10 = 90.$

II. $b * b = b^2 - 5b = 6$. So $b^2 - 5b - 6 = 0$.

$(b - 6)(b + 1) = 0$ $b = 6$ and $b = -1$.

NOTE
Many, many, many, many symbols are used beside
the *.

Let's do some practicing. But first. . . .

CHAPTER 20

PRACTICING FOR THE MATH SAT®

Congratulations on finishing the book. You are now ready to do the practice SATs. They are designed with questions that are similar to the kinds of questions on the real SATs. But there are some differences you should know.

1. The regular SATs are dull and boring. Mine are written so that you can have some fun with them. Hopefully, the day of the SAT you will remember these tests and laugh a little. Relaxing makes scores better.

2. If you notice, I don't put the formulas on the top of each section like the SAT. If you need to look, you will not do well. You must have the formulas memorized, you must.

3. If you notice, I haven't provided you with an SAT-like answer sheet. You won't use it anyway. Make sure you know how to use an SAT answer sheet.

4. Do not take the time limits too seriously. ETS, the maker of the SATs, has a very large staff and pretests all the questions. I am a staff of one.

5. The SATs have more reading-type questions (usually) than I do. Since they are one of a kind, I
   have not done as many because they probably
   won't do you much good. However, the SAT does
   give once in a lifetime surprise kinds of questions. I did put in a few of these.

6. Also, my tests may be a little harder than the real
   SATs since I have not put in as many easy questions as the SAT. Annnd, since the SAT pretests
   all of the problems, it can put them in order, easy
   to hard. I have tried, but I don't guarantee the
   order of difficulty.

Incidentally I have great respect for ETS. The hardest
thing I have evvvver written is a math SAT. It is so
hard you wouldn't believe!!!!

   Just like before, after each test or each part of each
test, read the solutions and make sure you understand
each problem. Good luck!!!!!

CHAPTER 21

PRACTICE TEST I

PART I

Use the empty places on these sheets to do all of your scratch work. When you decide which choice is correct, fill in letter A, B, C, D, or E on the answer sheet. On the real SAT, you will fill in a circle.

25 questions, 30 minutes.

1. $9 - x = x - 9$. $x =$

   A. −9
   B. 0
   C. 9
   D. 18
   E. 81

2. $x =$

   A. 30°
   B. 40°
   C. 70°
   D. 110°
   E. 290°

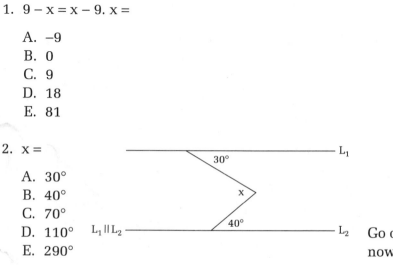

$L_1 \parallel L_2$

Go on to the next page now!!!! Do not hesitate.

3. $(3x^2)^2(2x^3)^3 =$

    A. $6x^{13}$
    B. $6x^{10}$
    C. $72x^{10}$
    D. $72x^{13}$
    E. $5x^{36}$

4. If Sandy paid $14 for 4 tickets, how much do 6 tickets cost?

    A. $7
    B. $14
    C. $21
    D. $28
    E. $56

5. y is 7 less than 3 times x. Then x =

    A. $(y + 7)/3$
    B. $(y - 7)/3$
    C. $3y + 7$
    D. $3y - 7$
    E. $7/3 - y/3$

6. If $x + 9$ is an even integer, the sum of the next two even integers is

    A. $x + 11$
    B. $2x + 22$
    C. $2x + 24$
    D. $3x + 39$
    E. $x^2 + 24x + 143$

7. If $\dfrac{3x}{4} = 5$, then $\dfrac{3x}{5} =$

    A. 20

    B. 20/3

    C. 3

    D. 4

    E. 5

8. If $a \pounds b = ab - b + b^2$, then $3 \pounds 5 =$

    A. 2

    B. 8

    C. 15

    D. 21

    E. 35

9. A board of length 45 feet is sawed into 2 pieces that are in the ratio of 2:3. The length of the larger piece is

    A. 9

    B. 18

    C. 27

    D. 36

    E. 45

10. 2000000000000000000000000000006 is exactly divisible by

    A. 2

    B. 3

    C. 9

    D. 2 and 3

    E. 2, 3 and 9

Go on to the next page.

11. $\sqrt{100 - 64} =$

  A. 2
  B. 3
  C. 4
  D. 6
  E. 8

12. If $x = -3$, then $-x^2 =$

  A. 6
  B. −6
  C. 9
  D. −9
  E. 81

13. x years in the future, Sandy will be y years old. z years in the future, Sandy will be how old?

  A. x + y + z
  B. x − y + z
  C. x − y − z
  D. x + z
  E. y − x + z

14. AB is parallel to CD annnd AD is parallel to BC. Angle A = 2x degrees. Angle C = 4x − 80 degrees. Angle B is how many degrees?

  A. 40°
  B. 80°
  C. 100°
  D. 110°
  E. 120°

15. $\dfrac{x}{y} = -2$ $x + y =$

    A. $-2y$
    B. 0
    C. $-y$
    D. $3y$
    E. $-3y$

16. 20 students received a 90 on a math test. 30 students got 100 on the same test. The class average (arithmetic mean) is

    A. 92
    B. 94
    C. 95
    D. 96
    E. 98

17. Let a, b, c, d, e, f, g, h be positive integers from 1 through 9, but we are not telling you what they are. a b c d 9 is exactly divisible by e f 7. The quotient could be

    A. g h 3
    B. g h 4
    C. g h 5
    D. g h 6
    E. g h 7

18. $10^{100} = 100^{x}.$ $x =$

    A. 50
    B. 100
    C. 110
    D. 1000
    E. 10

Go, go, go to the next page. Now!!!!

19. Given point P is on a line. A and B, not pictured, are on opposite sides of P such that 3AP = 4PB. M is the midpoint of AP.

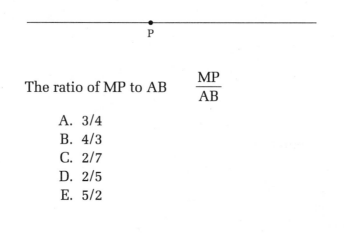

The ratio of MP to AB $\dfrac{MP}{AB}$

    A. 3/4
    B. 4/3
    C. 2/7
    D. 2/5
    E. 5/2

20. The sum of 5 consecutive integers is 105. The sum of the first 2 is

    A. 19
    B. 21
    C. 39
    D. 41
    E. 43

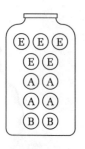

21. A jar contains 4 letter As, 2 letter Bs, and 5 letter Es. If a letter is pulled out at random, the probability of picking a vowel is

    A. 9/11
    B. 2/11
    C. 2/3
    D. 11/4
    E. 3/2

22. $\dfrac{\dfrac{4}{7}+\dfrac{1}{2}}{\dfrac{4}{7}-\dfrac{1}{2}}=$

    A. 15
    B. 1
    C. 0
    D. 5/3
    E. 3/5

23. $x = 2^{7n}.\ 16x =$

    A. 32^{7n}
    B. 2^{112n}
    C. 2^{7n+4}
    D. 2^{28n}
    E. 2^{11n}

24. $-1 < x < 0$. Arrange x^3, x^4, x^5 in order, smallest to largest.

    A. $x^3 < x^4 < x^5$
    B. $x^3 < x^5 < x^4$
    C. $x^5 < x^4 < x^3$
    D. $x^5 < x^3 < x^4$
    E. $x^4 < x^3 < x^5$

25. The sum of the 10 marked angles is

    A. 300°
    B. 360°
    C. 540°
    D. 600°
    E. 720°

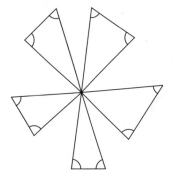

Stop, Stop! Stop!!!! Do NOT go on!!!!

PART 2

Select A if column A is larger.

Select B if column B is larger.

Select C if both columns are equal.

Select D if you cannot tell which column is larger or if they are equal.

25 questions, 30 minutes.

	Column A	Column B
1.	$5(3-5)$	$3(5-3)$
2. 0, 1, 2, 3, 0, . . .	87th term	2
3. $c+d=0$	c^2	d^2
4. $abc=0$ $\quad c \neq 0$	a	b
5.	$x-40$	100

6. (figure not drawn to scale)	m	n

	Column A	Column B
7. Boat A goes 1 mile in 4 minutes. Boat B goes 1 mile in 3 minutes.	speed of A	speed of B
8. a = the number of primes between 70 and 80. b = number of primes between 80 and 90.	a	b
9.	x	80

	Column A	Column B
10. $a^2 + 68a + 480 = 0$	$a^2 + 68a$	480
11. Car goes at 40 mph from C to D. Car goes from D back to C on the same path at 60 mph.	average speed	50

Go on to the next page.

Put A if column A is larger.

Put B if column B is larger.

Put C if the columns are equal.

Put D if you can't tell whether it is A or B or C.

	Column A	Column B
12.	m	p

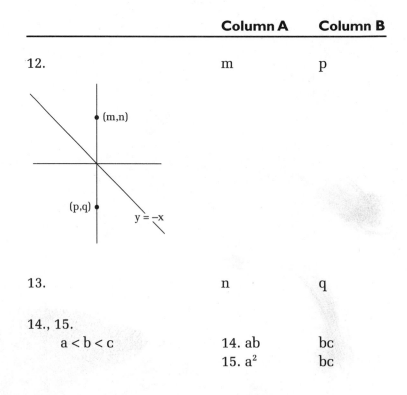

	Column A	Column B
13.	n	q

14., 15.

a < b < c

	Column A	Column B
14.	ab	bc
15.	a^2	bc

Use the blank spaces to do all scrap work. When you find the answer, write the answer on the answer sheet. (On the real SAT, you will fill in a grid.)

16. If $a^5 > a^3$ and $a > 0$, find a value for a that will make this inequality true.

17. The area of a circle is 36π. If half the circumference is $k\pi$, what is the value of k?

18. If $4 \leq n \leq 6$ $8 \leq m \leq 10$

    find the maximum value of $\dfrac{m + n}{m - n}$

19. $5b^2 = 25$. Find the value of $5b^6$.

20. The sum of the first 4 primes is

21. Find a fraction between 3/43 and 4/43. You may not have a decimal point in the answer. The denominator must be less than 100.

22. The distance between A and B is

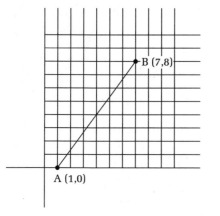

A (1,0)

B (7,8)

23. Given that there are 12 inches in a foot, the amount of cubic feet of dirt in a rectangular hole that is 6 ft by 6 ft by 9 in is

Go on to the next page. Immediately. At once. Don't delay. Right now!!!!

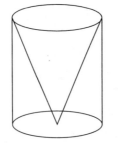

24. Given the volume of a cone is $(1/3)\pi r^2 h$. The volume of the cylinder pictured is 60. Find the volume of the cone pictured.

25. A monkey is trying to climb out of a 30-ft hole. In the morning, the monkey climbs up 3 feet. At night, the monkey is so tired he falls back 2 feet. The next day the same thing happens. The morning of what day does the monkey climb out of the hole?

Stop. Halt. Desist. Do
NOT go on.

PART 3

Use any empty space to do your scratchwork. Then decide which of the five answers is correct. Fill in on the line of the answer sheet that is found in this book.

10 questions, 15 minutes.

1. $x + 6 = 10$. Then $(x + 3)^2 =$
   A. 4
   B. 16
   C. 20
   D. 49
   E. 64

2. John can duplicate 2000 copies every 90 minutes. Mary can make 3000 copies every 90 minutes. How many copies can they make in 540 minutes if they alternate machines?

   A. 5,000
   B. 10,000
   C. 15,000
   D. 20,000
   E. 30,000

3. Which number is largest?

   A. .10101
   B. .101
   C. .1
   D. .1001
   E. .10011

Go on to the next page.

4. On the line segment connecting (5,6) and (21,18), which of the points below is on this line segment?

    A. (9,9)
    B. (12,12)
    C. (13,12)
    D. (12,13)
    E. (16,15)

5. How many times does the digit 5 appear on integers between 2 and 62?

    A. 5
    B. 6
    C. 15
    D. 16
    E. 17

6. If a mgms cost c cents, d mgms cost how many cents?

    A. ad/c
    B. a/dc
    C. dc/a
    D. c/ad
    E. acd

7. Given the list of integers 2, 3, 3, 3, 3, 5, 30 the median minus the mode is

    A. 0
    B. 1
    C. 2
    D. 3
    E. 28

8. $5x + 2y = 10$ $x + 8y = 46$ $3x + 5y =$

    A. 14
    B. 28
    C. 56
    D. 112
    E. Impossible to tell

9. $(x + 3)^2 = 6x + 25$

    I. $x = 0$

    II. $x = 4$

    III. $x = -4$ $x =$

    A. I only
    B. II only
    C. III only
    D. I and II
    E. II and III

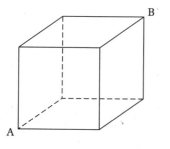

10. A cube pictured above has length of the edge 5. How many distinct paths along the edges from A to B has length exactly 15?

    A. 3
    B. 5
    C. 4
    D. 6
    E. 8

Stop!!!! Let's go to the answers.

ANSWER SHEET

Part I	Part 2	Part 3
1. _____	1. _____	1. _____
2. _____	2. _____	2. _____
3. _____	3. _____	3. _____
4. _____	4. _____	4. _____
5. _____	5. _____	5. _____
6. _____	6. _____	6. _____
7. _____	7. _____	7. _____
8. _____	8. _____	8. _____
9. _____	9. _____	9. _____
10. _____	10. _____	10. _____
11. _____	11. _____	
12. _____	12. _____	
13. _____	13. _____	
14. _____	14. _____	
15. _____	15. _____	
16. _____	16. _____	
17. _____	17. _____	
18. _____	18. _____	
19. _____	19. _____	
20. _____	20. _____	
21. _____	21. _____	
22. _____	22. _____	
23. _____	23. _____	
24. _____	24. _____	
25. _____	25. _____	

CHAPTER 22

PRACTICE TEST I
ANSWERS

PART I

1. Trial and error will show you the answer is 9 orrrr $-2x = -18$. So $x = 9$.   C

2. The secret is to draw the parallel line in the middle.

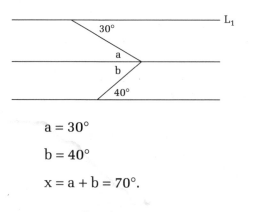

     $a = 30°$

     $b = 40°$

     $x = a + b = 70°$.   C

3. One way to do it is $(3x^2)(3x^2)(2x^3)(2x^3)(2x^3) = 72x^{13}$.

     D

4. $14 = 4$ tickets

  $ 7 = 2$ tickets

  $21 = 6$ tickets   C

5. Reading it we get $y = 3x - 7$. Remember, it is the equal sign annnnd less than reverses. Solving for x

  $3x - 7 = y$

  $3x = y + 7$

  $x = (y + 7)/3$ A

6. The space between even integers is allllways 2. $x + 11$ is the next; $x + 13$ is the next. The sum is $2x + 24$.   C

7. You could solve for x, buuut

  $$\frac{3x}{4} = \frac{5}{1}$$

  Interchange the 4 and 5, which you can do in a proportion

  $$\frac{3x}{5} = \ldots\ldots \frac{4}{1}$$ D

8. $a \pounds b = ab - b + b^2$. A "follow the rules" problem. $a = 3$; $b = 5$. $a \pounds b = 3(5) - 5 + 5^2 = 35$.   E

9. Easiest way to do it is $2x + 3x = 45$. $5x = 45$. $x = 9$. Larger piece is $3x = 3(9) = 27$.   C

10. There are two tricks (actually 3 involved). The number is divisible by 2 because the last digit, 6, is even. The sum of allll the digits is 8, which is not divisible by 3 or 9. Soooo, the answer is   A

11. Order of operations. $\sqrt{36} = 6$.   D

12. This is a problem I always give because so
     many people get it wrong, even advanced ones.
     $-x^2 = -(-3)(-3) = -9$ (3 minus signs is minus).   D

13. The trick is to put the age NOWWWW. y in x
     years in the future? $y - x$ now. $y - x + z$,
     z years in the future.   E

14. A = C. $2x = 4x - 80$. $x = 40°$. Sooo, A = 80°.
     B is sup = 100°.   C

15. $\dfrac{x}{y} = \dfrac{-2}{1}$ $x = -2y$ $x + y = -2y + y = -y$ C

16. $20(90) = 1800$ points; $30(100) = 3000$ points.
     Total, 4800 points divided by 50 = 96.   D

17. Why does $486/27 = 18$? Because $27 \times 18 = 486$.
     More specifically, the last digits must multiply.
     The answer is E, since (gh7)(ef7) must end in
     a 9 since $7 \times 7 = 49$. The others won't.   E

18. $10^{100} = (10^2)^x$. $2x = 100$. Sooo, $x = 50$.   A

19. $3\ AP = 4\ PB$. Let PB = 3 units and AP = 4 units
     and M is located as pictured. $MP/AB = 2/7$,
     just count.   C

20. If it is an odd number, the middle is $105/5 =$
     21. First 2 are before 19 and 20. The sum is 39.   C

21. 9 vowels, 2 consonants. 9/11.   A

22. LCD 14

$$\frac{\dfrac{14}{1}\dfrac{4}{7}+\dfrac{14}{1}\dfrac{1}{2}}{\dfrac{14}{1}\dfrac{4}{7}-\dfrac{14}{1}\dfrac{1}{2}}=\frac{8+7}{8-7}=\frac{15}{1}=15$$
A

23. $16 = 2^4$. $16x = 2^4 2^{7n} = 2^{7n+4}$, adding the exponents.
C

24. x^4 must be biggest because it is the only positive. Let $x = -\frac{1}{2}$.

$(-\frac{1}{2})^3 = -1/8$ $(-\frac{1}{2})^5 = -1/32$ bigger than $-1/8$.

Remember negatives reverse. Still don't believe? Look!
B

25. The sum of the 60° angles, 6 of them, is 360°. Remember the trick.
B

PART 2

1. $A = 5(-2) = -10$. B is positive.
B

2. The cycle is 4. $4\overline{)87}$. We don't care about the answer. The remainder is 3. The third term is 2.
C

3. $c = -d$. $(c)^2 = (-d)^2$. $c^2 = d^2$.
C

4. c is not 0. So a or b is 0, but we don't know which one. (Could be both.)
D

5. $x + 40 = 180$. $x = 140$. $140 - 40 = 100$.
C

6. Larger side lies opposite larger angle. n is
    opposite 10°.   A

7. Rate = distance over time. A = 1/4, B = 1/3.
    A is faster, but B is bigger; trickkkky.   B

8. a = 71, 73, 79 (3 numbers). b = 83, 89
    (2 numbers). a is bigger.   A

9. $2x + 10 = 180$. $x = 85$.   A

10. Fake factoring. If $a^2 + 68a + 480 = 0$, $a^2 + 68a$
    must $= -480$.   B

11. Toughie. Because the distance does not matter,
    let d = 120 since 40 and 60 divide into it with
    no remainder. 120/40 = 3 hr. 120/60 = 2 hr.
    Total time 5 hr. Total distance = 240 miles/
    5 = 48 mph. The average speed is NOT NOT NOT
    the average of the speeds.   B

12. . . . On the y axis, the x coordinates are
    both 0.   (12.) C

13. $n > 0$ and $q < 0$. ($y = -x$ is drawn to con-
    fuse you.)   (13.) A

14. If b = 0, they both are equal. $-5 < -2 < 3$
    ab $> bc$.   D

15. $b = 0$ $a^2 > bc$ $2 < 3 < 4$ $a^2 < bc$ D

16. An easy one; any number bigger than 1 is ok.

17. $\pi r^2 = 36\pi$. So r = 6. c = $2\pi r$ = 12π. Half of this
    is 6π. k = 6.

18. A real toughie. Doesn't fit the patterns we have learned. The max and min values occur at the extremes. So n = 4 or 6, m = 8 or 10.

$$\frac{10 + 4}{10 - 4} = 14/6 \qquad \frac{10 + 6}{10 - 6} = 16/4 = 4 \qquad \frac{8 + 4}{8 - 4} = 4$$

$$\frac{8 + 6}{8 - 6} = 14/2 = 7. \qquad \text{The answer is 7.}$$

19. $5b^6 = 5b^2(b^2)(b^2)$. $5b^2 = 25$. Sooooo, $b^2 = 5$. $25(5)(5) = 625$.

20. $2 + 3 + 5 + 7 = 17$. 1 is not a prime. 2 is the only even prime.

21. $\dfrac{3}{43} = \dfrac{6}{86} \qquad \dfrac{4}{43} = \dfrac{8}{86}.$ Between is 7/86.

22. Straight distance formula $\sqrt{(7 - 1)^2 + (8 - 0)^2} = 10$.

23. 0. There is no dirt in a hole. Some of the SAT is just fun.

24. Much easier than it looks if you see the trick. You don't need to know r or h. Good thing. You can't find them. All you need to know is the volume of a cone is 1/3 the cylinder since h and r are the same. 1/3 of 60 is 20.

25. After 27 days, the monkey has netted 27 feet. On the 28th morning, the monkey climbs 3 ft. 27 + 3 = 30. The monkey is out of the hole. 28.

PART 3

1. $x = 4$. $x + 3 = 7$. $7^2 = 49$.   D

2. The trick is alternate. You must read this
   word.

   $2000 + 3000 + 2000 + 3000 + 2000 + 3000 = 15,000$.   C

3.

   A .10101
   B .10100
   C .10000
   D .10010
   E .10011

   First 2 digits are the same, for all. Third digit eliminates C, D, E. Fifth digit eliminates B.   A

4. $m = (18 - 6)/(21 - 5) = 12/15 = 3/4$. Only point
   that has same slope with (5,6) is   A

5. 5, 15, 25, 35, 45, 50, 51, 52, 53, 54, 55 (2 5s),
   56, 57, 58, 59   D

6. $\dfrac{a}{c} = \dfrac{x}{d}$. $x = dc/a$ C

7. Median is middle, 3. Mode is most common,
   3. $3 - 3 = 0$.   A

8. Add. $6x + 10y = 56$. Divide by 2. $3x + 5y = 28$.   B

9. Trial and error, orrr $x^2 + 6x + 9 = 6x + 25$.
   $x^2 = 16$. $x = \pm 4$.   E

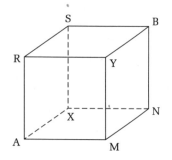

10. AMNB, AMYB, ARYB, ARSB, AXSB,
     AXNB.   6   D

Let's try another. You have no idea how hard it is
to make one of these up. Don't tell me it's harder
to take it. I think these are fun to take. Hopefully,
after you read this book, you will not only do this
book muuuuch better but enjoy doing these
problems. Let's go on.

CHAPTER 23

PRACTICE TEST II

PART I

Select the best answer, A, B, C, D, or E, and put it on your form. On the real SAT, you will fill in a circle or something like a circle.

25 questions, 30 minutes.

1. $9 - 5x = 5.$ $3 - 5x =$

    A. -1
    B. 0
    C. 5
    D. 6
    E. 9

2. $x =$

    A. 30
    B. 50
    C. 55
    D. 65
    E. 110

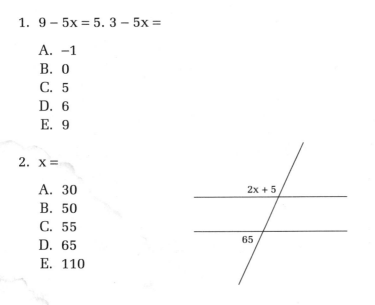

Go immmmediately to the next page!!!!

3. $\dfrac{x}{6} + \dfrac{x}{12} = 2. \; x =$

   A. 2
   B. 4
   C. 6
   D. 8
   E. 12

4. $x - y = 5. \; xy = 24. \; x + y$ could be

   A. 9
   B. 11
   C. 14
   D. 17
   E. 20

5. 1, 3, 6, 10, 15, . . . What is the 10th term of the sequence?

   A. 40
   B. 50
   C. 55
   D. 61
   E. 72

6. 2 bats and 2 balls cost $48.00. 7 bats and 6 balls cost $182.00. One ball and one bat cost. . . .

   A. $18
   B. $22
   C. $24
   D. $28
   E. $32

7. $\dfrac{x\,(x^{a+b})}{x^b} =$

    A. $(x^2)^{a+2b}$

    B. x^{2a}

    C. x^{a+2b+1}

    D. x^{a+1}

    E. x^{ab+b^2}

8. $7x + 5y = 39$. $5x + 3y = 11$. What is $(x + y)/2$?

    A. 7

    B. 14

    C. 18

    D. 21

    E. 28

9. $2^{60} = 8^{4x}$. $x =$

    A. 15

    B. 7.5

    C. 5

    D. $2\frac{1}{3}$

    E. 20

10. $(-1)^2 + (-1)^3 + (-1)^4 + (-1)^5 + (-1)^6 + (-1)^7 + (-1)^8 + (-1)^9 + (-1)^{10}$

    A. -9

    B. -1

    C. 0

    D. 1

    E. 9

Go, go, go to the next page.

11., 12.

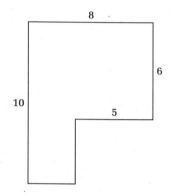

11. The perimeter of the figure is

    A. 29
    B. 58
    C. 36
    D. 54
    E. 72

12. The area of this figure is

    A. 80
    B. 68
    C. 60
    D. 50
    E. 40

13. to 15.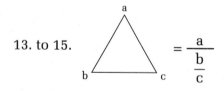

13.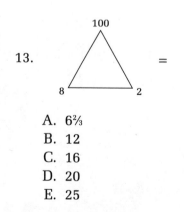

    A. 6⅔
    B. 12
    C. 16
    D. 20
    E. 25

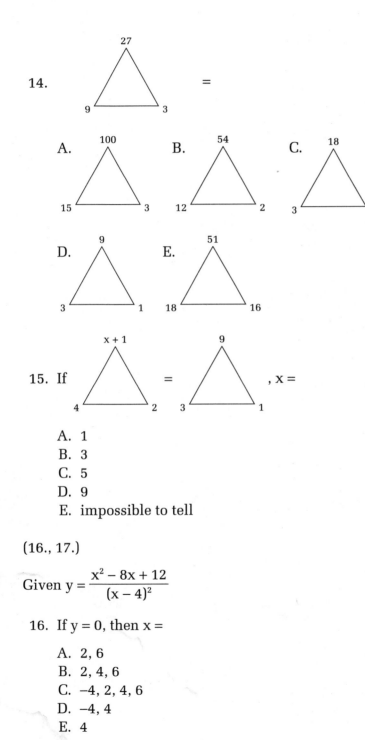

14. (triangle with 27, 9, 3) =

A. (triangle with 100, 15, 3)   B. (triangle with 54, 12, 2)   C. (triangle with 18, 3, 3)

D. (triangle with 9, 3, 1)   E. (triangle with 51, 18, 16)

15. If (triangle with $x + 1$, 4, 2) = (triangle with 9, 3, 1), $x =$

A. 1
B. 3
C. 5
D. 9
E. impossible to tell

(16., 17.)

Given $y = \dfrac{x^2 - 8x + 12}{(x - 4)^2}$

16. If $y = 0$, then $x =$

A. 2, 6
B. 2, 4, 6
C. −4, 2, 4, 6
D. −4, 4
E. 4

Pretty please, go to the next page.

17. For what value(s) of x does y have no value?

    A. 2, 6
    B. 2, 4, 6
    C. −4, 2, 4, 6
    D. −4, 4
    E. 4

18. Each brother has an equal number of sisters. Each sister has twice as many brothers. How many brothers and sisters are there?

    A. 2 brothers, 2 sisters
    B. 3 brothers, 3 sisters
    C. 3 brothers, 2 sisters
    D. 4 brothers, 3 sisters
    E. 6 brothers, 4 sisters

19. The volume of a cube is 64. Its surface area is

    A. 4
    B. 8
    C. 16
    D. 48
    E. 96

20. If this same cube has its faces colored, what is the minimum number of colors so that no two surfaces that touch have the same color?

    A. 1
    B. 2
    C. 3
    D. 4
    E. 5

21. The sum of measures of a triangle are consecutive integers. The largest is?

    A. 58
    B. 59
    C. 60
    D. 61
    E. 180

22. The fewest number of trees needed for 6 rows of 4 apple trees each is

    A. 24
    B. 23
    C. 20
    D. 19
    E. 12

23. An $875.20 TV is reduced in price by 50%. What % increase is needed to restore it to $875.20?

    A. 50%
    B. 75%
    C. 100%
    D. 150%
    E. 200%

24. The three semicircles are as pictured. Find their total area.

    A. 12.5π
    B. 25π
    C. 50π
    D. 75π
    E. 100π

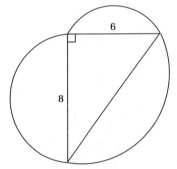

25. $2^n + 2^n =$

    A. 2^{n^2}
    B. 2^{2n}
    C. 4^{n^2}
    D. $(2 + 2)^{n^2}$
    E. 2^{n+1}

Stopppp nowwww!!

PART 2

Put A if column A is bigger.

Put B if column B is bigger.

Put C if the columns are equal.

Put D if you cannot tell which is bigger or if they are equal.

25 questions, 35 minutes.

	Column A	Column B
1. $x^2 = 25$ $y^2 = 100$	x	y
2.	$.3a + .5b$	$.5(a + b)$
3. $a, b > 2$	$\dfrac{a}{b}$	$\dfrac{a - 2}{b - 2}$
4. $b > 0$	$\sqrt{b + 9}$	$\sqrt{b} + 3$
5. Let S = the sum of the roots of the equation $x^3 - 2x^2 - 8x = 0$	S	0
6. $-40 < x < -2$	$1/x^8$	$1/x^5$

	Column A	Column B

7. $A = \dfrac{48765 \times 852 \times 361 \times 8 \times 10}{2 \times 3 \times 4}$

   $B = \dfrac{7 \times 9 \times 852 \times 361 \times 48765}{3 \times 4 \times 5}$ A   B

8. ABCD is a square.   Area of   18
   AB = 6.   shaded
                         portion

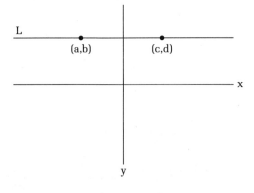

9. $a \circ b = b/2 + a$ $4 \circ 8$ 8

10., 11.

L is perpendicular to the
y axis.

10.   a   c

11.   b   d

12. $4 \leq x, y \leq 8$ x/y y^2/x

13. $x < -1$, x integer $x^5 - 3$ 0

Go on to the next page;
yes, please do.

14. $\sqrt{x}/x$ $1/\sqrt{x}$

15. x, y are integers.
     $x^2 < 65$
     $y^4 < 82$ minimum   minimum
                                   value of x   value of y

Solve the following problems and put the answers on your answer sheet. On the real SAT, you will fill in the gridddd.

16. How many rectangles 2 inches by 3 inches, can be cut from a piece of paper that is 9 inches by 1 foot? 12 inches = 1 foot.

17. 15 less than 4 times a number is 5 more than twice the number. What is the number?

18. The perimeter of an equilateral triangle is exactly the same as the perimeter of an octagon. Find the ratio of the side of the triangle to the side of the octagon.

19. Eight people are standing in a circle. They shake each other's hands. How many handshakes if you count each pair shaking once and only once?

20. An octagon has how many diagonals?

21. A team has won 12 out of 20 games. How many games does the team have to win in a row for their winning percentage to be 75%?

22. If $1 < x < 50$, find the number of even numbers that are NOT multiples of 3.

23. Find the area.

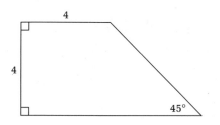

24. A rectangular box with a square base has a volume the same as a cylinder. One side of the base is $\sqrt{8\pi}$. The height of the box is 6. If the height of the cylinder is 3, find the radius of the base.

25. There are two numbers x such that the distance between (x,3) and (2,7) is 5. Find the sum of these numbers.

Stop!!!! This is the end
of this section!!!!

PART 3

Select your best choice, A, B, C, D, E and put it on your answer page. On the real SAT, you will fill in something.

10 questions, 15 minutes.

1. J.D. wants to get an A in math. A 90 average is needed. J.D. has grades of 87, 79, and 96 on the first 3 tests. What does J.D. need for an A average?

   A. 88
   B. 92
   C. 96
   D. 98
   E. An A is not possible

2. After a 20% discount, a radio costs $56. What was the original cost?

   A. $44.80
   B. $67.20
   C. $70.00
   D. $76.00
   E. $80.00

3. At 10:25, a car left a location. 20 miles later at 10:40, the car arrived. The speed of the car is

   A. 40 mph
   B. 55 mph
   C. 60 mph
   D. 80 mph
   E. 132⅔ mph

4. United States postage costs 32¢ for the first ounce and 23¢ for each additional ounce. Find the cost of mailing a ½ pound letter.

   A. $1.93
   B. $2.16
   C. $2.17
   D. $2.50
   E. $2.52

5. △ABC and △BDQ are 2 ≅ isosceles right triangle. The coordinates of point Q are

   A. (10,0)
   B. (10,–10)
   C. (10,–20)
   D. (20,–10)
   E. (20,–20)

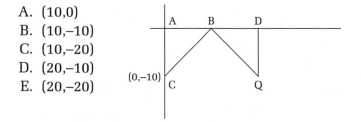

In the next 5 questions, each one will have 2 statements, 1 and 2. Decide if each is true or false (if a statement is sometimes true, it is false).

   Put A if 1 is false and 2 is false.

   Put B if 1 is false and 2 is true.

   Put C if 1 is true and 2 is false.

   Put D if 1 is true and 2 is true, but they are not related.

   Put E. if 1 is true, and 2 is true, and they are related.

6.

   1. If 2 parallel lines are cut by a transversal, the measure of all acute angles is congruent.

   2. The sum of the measures of the angles of a polygon is 360°.

7.

    1. If $3x + 5 = 20$, then $x = 5$.

    2. If $2y + 7 = 15$, then $y = 4$.

8.

    1. The vertex angle of an isosceles triangle is 40°. One base angle is 70°.

    2. In a scalene triangle, either two angles may be equal or not.

9.

    1. The area of a rectangle is length times width.

    2. The area of a triangle is ½ base times height.

10.

    1. If you double the radius of a circle, you double its area.

    2. If you triple the edge of a cube, the volume increases by 6 times.

Stop, stop, stop. I'm sure you want to. Let's check the answers.

ANSWER SHEET

Part 1	Part 2	Part 3
1. _____	1. _____	1. _____
2. _____	2. _____	2. _____
3. _____	3. _____	3. _____
4. _____	4. _____	4. _____
5. _____	5. _____	5. _____
6. _____	6. _____	6. _____
7. _____	7. _____	7. _____
8. _____	8. _____	8. _____
9. _____	9. _____	9. _____
10. _____	10. _____	10. _____
11. _____	11. _____	
12. _____	12. _____	
13. _____	13. _____	
14. _____	14. _____	
15. _____	15. _____	
16. _____	16. _____	
17. _____	17. _____	
18. _____	18. _____	
19. _____	19. _____	
20. _____	20. _____	
21. _____	21. _____	
22. _____	22. _____	
23. _____	23. _____	
24. _____	24. _____	
25. _____	25. _____	

CHAPTER 24

PRACTICE TEST II ANSWERS

PART I

1. $3 - 5x$ is 6 less than $9 - 5x$. $5 - 6 = -1$.   A

2. $2x + 5 + 65 = 180$. $2x = 110$. $x = 55$.   C

3. Multiply by 12. $2x + x = 24$. $x = 8$.   D

4. This is trial and error. $x = 8$, $y = 3$; $x + y = 11$.   B

5. Just count. . . . 21, 28, 36, 45, 55.   C

6. This is a real reading trick. You do not need the second equation at all. 2 bats plus 2 balls is $48. Soooooo, one of each is $24.   C

7. $x^1 x^{a+b} x^{-b} = x^{a+1}$.   D

8. When I did this the first time, my arithmetic wasn't too good. I hope yours is better.

Subtracting, we get $2x + 2y = 28$.

Soooo $\qquad\qquad x + y = 14$.

Annnnd $\qquad\qquad \dfrac{x + y}{2} = 7$.

9. $2^{60} = (2^3)^{4x}$. $12x = 60$. $x = 5$. $\qquad$ C

10. $1 - 1 + 1 - 1 + 1 - 1 + 1 - 1 + 1 = 0 + 0 + 0 + 0 + 1 = 1!$ $\qquad$ D

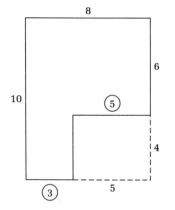

11., 12.

$3 + 5 = 8$.

$6 + 4 = 10$.

$P = 10 + 8 + 6 + 4 + 5 + 3 = 36$. $\qquad$ C

$A = 8 \times 10 - 4 \times 5 = 80 - 20 = 60$. $\qquad$ C

13. $\dfrac{100}{\frac{8}{2}} = \dfrac{100}{4} = 25$. $\qquad$ E

14. $\dfrac{27}{\frac{9}{3}} = \dfrac{27}{3} = 9$. $\qquad$ $\dfrac{54}{\frac{12}{2}} = \dfrac{54}{6} = 9$. $\qquad$ B

15. $\dfrac{x+1}{\frac{4}{2}} = \dfrac{9}{\frac{3}{1}}$ $\qquad$ $\dfrac{x+1}{2} = 3$ $\qquad$ $x + 1 = 6$ $\qquad$ $x = 5$ $\qquad$ C

16., 17. $y = \dfrac{(x-6)(x-2)}{(x-4)(x-4)}$ $y = 0$ top = 0

   no value bottom = 0

(16.)   A

(17.)   E

18. Each brother has 3 brothers and 3 sisters.
     Each sister has 4 brothers and 2 sisters.   D

19. $e^3 = 64.$ e = 4. $6e^2 = 6(16) = 96.$ E

20.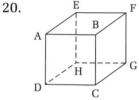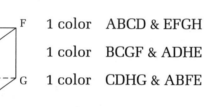

     1 color   ABCD & EFGH

     1 color   BCGF & ADHE

     1 color   CDHG & ABFE

                     3 colors   C

21. $x + x + 1 + x + 2 = 180.$ x = 59. x + 2 = 61.   D

22. Trick   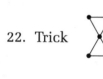   12, count.   E

23. Trick. Amount doesn't matter. $100 to $50.
     50% dip. $50 to $100. 100% increase.   C

24. Hypotenuse is 10, of course, thanks to old Pythag.
     A trick that would be nice to know to save time.
     We know $c^2 = a^2 + b^2$ (squares). It is also true for
     circles and semicircles. The two smaller semi-
     circles add to the larger one. We only have
     to find the areas of the 2 largest semicircles,
     one full circle radius 5 (diameter is 10).   B

25. We did this one. $2^n + 2^n = 1(2^n) + 1(2^n) = 2(2^n) = 2^1 2^n = 2^{n+1}$. E

PART 2

1. $x = \pm 5$, $y = \pm 10$ D

2. Trick. .5a and .3a depend on a positive or negative. D

3. We don't know if a or b is bigger. D

4. We talked about this. (2 square roots always bigger.) B

5. $x(x - 4)(x + 2) = 0$ $x = 0, 4, -2$ Sum is 2. A

6. All we need to know is that x is a negative. (A is positive; B is negative.) A

7. We don't do the arithmetic. We don't care what the answer is, only which is bigger. Cancel the 48765, 852, 361, 3, 4 from both sides.

$$\left(\text{since } \frac{8 \times 10}{2} > \frac{7 \times 9}{5} \right)$$ A

8. Area of unshaded portion is ½ the square. So is the unshaded portion. C

9. $(8/2) + 4 = 8$. C

10., 11. y = same y value $a < 0$ $b > 0$

(11.) C

(10.) B

12. Toughie. $x = 8$
    $y = 4$

    $\dfrac{8}{4} = 2$ $\dfrac{y^2}{x} = \dfrac{16}{8} = 2$. Otherwise y^2/x is bigger.   D

13. $(\text{neg})^5 = \text{neg}$ $\text{neg} - 3 = \text{neg} < 0$ B

14. $\dfrac{\sqrt{x}}{x} \cdot \dfrac{\sqrt{x}}{\sqrt{x}} = \dfrac{x}{x\sqrt{x}} = \dfrac{1}{\sqrt{x}}$ C

15. TRICK.

    $x^2 < 64$ smallest $\boxed{x = -8}$
    $y^4 < 82$ smallest $y = -3$ B

16. $2 \times 3 = 6$ $9 \times 12 = 108$ $108 \div 6 =$ 18

17. $4n - 15 = 2n + 5$
    $2n = 20$
    $n =$ 10

18. $3s = 8\theta$ $s/\theta =$ 8/3

19. Each person shakes 7 hands, but each is
    counted twice. Sooooo (8)(7)/2   28

20. Looks the same but isn't. Each vertex can't be
    drawn to itself or to either side. 5 total are left.
    Each is counted twice. $5(8)/2 = 20$. There is a
    formula if you like. The number of diagonals
    is $n(n - 3)/2$. $n \geq 3$.   20

21. $\dfrac{12 + x}{20 + x} = \dfrac{3}{4}$ (75%) $4x + 48 = 3x + 60$ $x =$ 12

22. 2 into 48 is 24. 6 into 48 is 8 (mult of 6 are
    both 2 and 3). $24 - 8 =$ 16

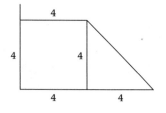

23. Area = trapezoid or square + triangle =
    16 + 8 =   24

24. $V = s \times s \times h = \sqrt{8\pi} \times \sqrt{8\pi} \times 6 = 48\pi.$ $V =$
    $\pi r^2 h = 48\pi.$ $h = 3.$ $\pi r^2 = 16\pi.$ $r^2 = 16.$ $r =$ 4

25. $(x-2)^2 + (7-3)^2 = 25.$ $x^2 - 4x + 4 + 16 = 25.$
    $x^2 - 4x - 5 = 0.$ $(x-5)(x+1) = 0.$ $x = 5.$
    $x = -1.$ $5 + -1 =$ 4

PART 3

1. $87 + 79 + 96 = 262.$ $90 \times 4 = 360.$ $360 - 262 =$
   98. Tough A, but possible.   D

2. You've got to know percents. $.8x = 56.$ $x =$
   $\$70 = \dfrac{560}{8}.$ C

3. 15 minutes. 1 little quarter of an hour.
   $4 \times 20 = 80$ mph.   D

4. $32\text{¢} + 7$ (not 8) $\times 23\text{¢} =$ A

5.   D

6. 1 is true. 2 is false (should be quadrilateral).   C

7. Both are true, but certainly unrelated.   D

8.

  1. $40 + 2x = 180$. $2x = 140$. $x = 70$. True.

  2. Scalene all sides not equal means all angles
      not equal—false.   C

9. A triangle is 1/2 a rectangle. I hope you've
    been told that. Both true and related.   E

10. $r = 10\pi$. $A = 100\pi$. $r = 20$. $A = 400\pi$. False
     (4 times; not 2). $e = 2$. $e^3 = 8$. $e = 6$. $e^3 = 216$.
     $216/8 = 27$ times, not 6. False.   A

Let's take a deep breath and try one more test.

CHAPTER 25

PRACTICE TEST III

PART I

25 questions, 30 minutes.

Put A if column A is bigger.

B if column B is bigger.

C if the columns are equal.

D if you can't tell.

	Column A	Column B
1. $x^2 = 9$, $y^2 = 9$	x	y
2.	.5%	5/100
3. A rug costs $10 a square foot. The rug is 2 yd by 3 yd.	cost of the rug	$500
4. $x^2 - x - 6 = 0$	x	4

Go, go, go on to the next page.

Put A if column A is bigger.
B if column B is bigger.
C if the columns are equal.
D if you can't tell.

		Column A	**Column B**
5.	$2 \leq x \leq 6$	$\left(\dfrac{x}{.08}\right)^2$	$x/8$
6.		$(a - b)^2$	$a^2 - b^2$
7.	26% of x = 88. 52% of y = 44.	x	y

8.

$\sphericalangle C > 90°$
Diagram is not drawn to scale.

		Column A	**Column B**
8.		$a^2 + b^2$	c^2
9.	$-10 \leq a \leq -7$	$1/a$	$2a$
10.	$x < 0$	$a + b - x$	$x + b + a$
11., 12.	$47 < a < b < c < 78$		
(11.)		ab	bc
(12.)		ac	b^2
13.		$0/a - 0(a) - 0$	0
14.	$a \neq 0$	$(a/4)^3$	$a^3/4$

	Column A	Column B

15.

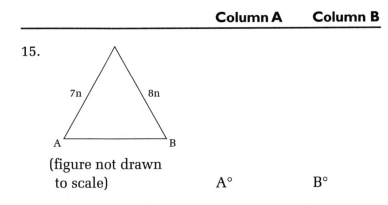

(figure not drawn
to scale)   A°   B°

Select the correct answer and put it on the line or just
circle it in the book. The SAT will have you fill in
something.

16. $\dfrac{(9-x)^2}{2} = \dfrac{(x-3)^2}{2}$ x =

   A. 0
   B. 3
   C. 4
   D. 6
   E. 8

17. $\left(\dfrac{a^4b^7}{a^6b^4}\right)^3 =$

   A. $\dfrac{b^9}{a^6}$

   B. $\dfrac{b^{281}}{a^{256}}$

   C. $\dfrac{b^3}{a^2}$

   D. a^6b^{17}
   E. $a^{58}b^{339}$ Go rapidly to the next
page.

18. $\left(\dfrac{6 \times 6 \times 6}{6 + 6 + 6}\right) y = 6.$ y =

    A. 1/6
    B. 1/2
    C. 6
    D. 36
    E. 216

19. $\dfrac{80 + 2x}{8}$ is equivalent to

    A. 10 + 2x
    B. 80 + x/4
    C. (80 + x)/4
    D. (10 + x)/4
    E. $10 + \dfrac{x}{4}$

20. $\dfrac{x^2 + 9x + 8}{x^2 + 4x + 4} = 1.$ x =

    A. −17/8
    B. −4/5
    C. −2
    D. −1
    E. −8

21. ABCD is a square. BP = 6. The area is

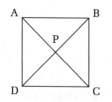

    A. 144
    B. $72\sqrt{2}$
    C. 72
    D. $36\sqrt{2}$
    E. 36

22. Speaking of squares, if you add 3″ to the length and subtract 3″ from the width, the area of the new rectangle is

    A. 9 square inches less than the square
    B. 3 square inches less than the square
    C. The same as the square
    D. 3 square inches more than the square
    E. 9 square inches more than the square

23. $\left(x - \dfrac{1}{x}\right)^2 = 100 \qquad x^2 + \dfrac{1}{x^2} =$

    A. 8
    B. 12
    C. 98
    D. 102
    E. 10,000

24. Which letter indicates a 50% increase from 1 to 2 followed by a 50% decrease from 2 to 3?

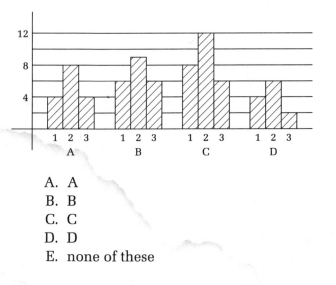

    A. A
    B. B
    C. C
    D. D
    E. none of these

Go rapidly to the next page.

25. A clock rings at equal intervals forever: ding, ding, ding, ding, ding, . . . The bell rings 6 times in 5 seconds. In 30 seconds the bell rings how many times?

　　A.　30
　　B.　31
　　C.　36
　　D.　40
　　E.　61

It is time to stop for a moment. STOP!!!!

PART 2

Select the correct answer and fill it in the box. The SAT does it differently.

25 questions, 25 minutes.

1.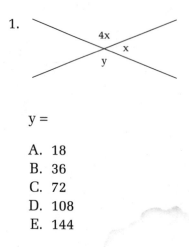

　　y =

　　A.　18
　　B.　36
　　C.　72
　　D.　108
　　E.　144

2. $x^4 - 17 = 64$. $x^4 =$

    A. 81
    B. 9
    C. ± 9
    D. 3
    E. ± 3

3. $(10ab^2)^3 =$

    A. $30ab^5$
    B. $30ab^6$
    C. $1000ab^6$
    D. $1000a^3b^5$
    E. $1000a^3b^6$

4. 4 zoups = 7 zims annnnd 8 zims = 11 zounds. The ratio of zoups/zounds is

    A. 77/32
    B. 32/77
    C. 7/22
    D. 22/7
    E. 2

5. The value of q quarters and d dimes is

    A. $d + q$
    B. $10d + q$
    C. $10(d + 25q)$
    D. $25(d + q)$
    E. $10d + 25q$

Go, go go on to the next page.

6.

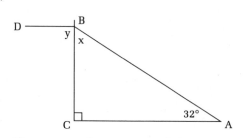

(figure not drawn to scale)

AC is parallel to BD. $x + y =$

A. 90
B. 122
C. 132
D. 138
E. 148

7. $\dfrac{\frac{3}{4}}{3m} =$

A. m/4
B. 4/m
C. 36/m
D. m/36
E. 9m/4

8. If $x = 10$, $(10x)^2 + 10x^2 =$

A. 10,100
B. 11,000
C. 20,000
D. 100,000
E. 10,000,000

9., 10. $a * b = 2ab + b$. $4 * 5 =$

    A. 44
    B. 45
    C. 160
    D. 200
    E. 9

10. If $a * b = 0$ and $b \neq 0$, then $a =$

    A. $-\frac{1}{4}$
    B. $-\frac{1}{2}$
    C. 0
    D. 1
    E. 2

11. The equation that best illustrates the chart is

    A. $y = 2x^2 - 4$
    B. $y = x^2$
    C. $y = 3x - 2$
    D. $y = x + 3$
    E. $y = 2x - 1$

x	1	2	3	4
y	1	4	7	10

12. k/2 odd and k/5 even. k could be

    A. 15
    B. 20
    C. 30
    D. 40
    E. 60

Let's journey to the
next page.

13. $x = \dfrac{ay}{b} + c$. Then $y =$

    A. $abc + c$
    B. $(bx - c)/a$
    C. $(bx + c)/a$
    D. $(ab - ac)/x$
    E. $(bx - bc)/a$

14. O center of circle. $OB = 5$, $BC = 2$. The area of the shaded region is

    A. 75π
    B. 21π
    C. 24π
    D. 74π
    E. 32π

15. O center of the circle. Radius 10. The area of the shaded part is

    A. $25(\pi - 1)$
    B. $25(\pi - 2)$
    C. $100(\pi - \frac{1}{4})$
    D. $50(\pi - 1)$
    E. $50(\pi - 2)$

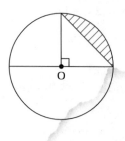

16. For the line $4x + 5y = 20$, the slope is

    A. 20
    B. −5
    C. −4
    D. −4/5
    E. −5/4

17. $10^2 − (−8)^2 =$

    A. 36
    B. −36
    C. 164
    D. 6400
    E. 4

18. (figure not drawn to scale) The ratio of angle A to angle B is 3:2. Angle A =

    A. 30
    B. 60
    C. 75
    D. 90
    E. 120

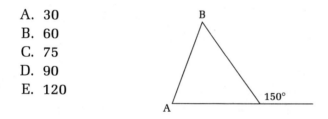

19. If $y = x^2/z^3$, then $xy/z =$

    A. x^3/z^4
    B. x/z^2
    C. x^3/z^3
    D. x/z
    E. xz

Please continue to next page.

20. $9^{3x+3} = 27^{4x-5}$ $x =$

  A. 8
  B. 5
  C. 7/2
  D. 5/2
  E. 3

21. A basketball team has 4 players whose heights were 6'7", 6'9", 6'10", 7'1". How tall must the 5th player be so the average (arithmetic mean) is 7'?

  A. 7'3"
  B. 7'6"
  C. 7'9"
  D. 7'11"
  E. 8'

22. ABCD is a square. AB = 1. The area of the entire region is

  A. $1 + \pi$
  B. $1 + \pi/2$
  C. $1 + \pi/4$
  D. $4 + \pi$
  E. $4 + \pi/2$

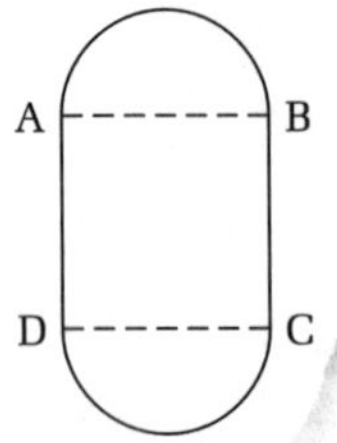

23. O center of the circle. CO = 6. Area of shaded
    part = 24π. ∡x =

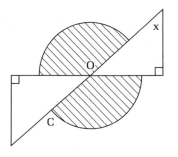

    A. 15°
    B. 30°
    C. 45°
    D. 60°
    E. 100°

24. x =

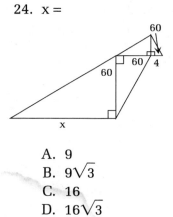

    A. 9
    B. $9\sqrt{3}$
    C. 16
    D. $16\sqrt{3}$
    E. 36

Shhhh. Quietly but
quickly go on to the
next page.

25.

	year 5760		year 6000
	32,000 glzrs		80,000 glzrs

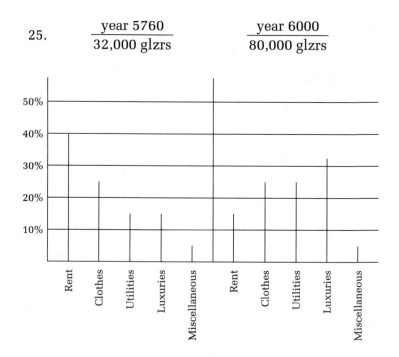

On planet zyzxz, in the year 5760, the budget was 32,000 glzrs. In the year 6000, the budget was 80,000 glzrs. Which part of the budget, approximately, was the same number of glzrs?

A. rent
B. clothes
C. utilities
D. luxuries
E. miscellaneous

Stopppp!!!!

PART 3

Find the answer to each question and put it on the answer sheet. The SAT will have you fill in a silly grid.

10 questions, 15 minutes.

1. Find the sum of all the primes between 50 and 60.

2. The product of two consecutive even positive integers is 48. The quotient of the larger divided by the smaller is . . .

3. Find a fraction between 7/8 and 7/9. There may be no decimal in the answer and the denominator must be less than 100.

4. The sum of 15 consecutive integers is 0. Find the product of the two smallest.

5. $4^{n+1} = 64$. $2^{n+2} + 3^{n+1} + 5^{n-1} =$

6.

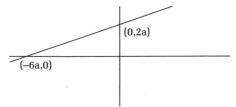

$(0, 2a)$

$(-6a, 0)$

The slope of this line is what?

7. The supplement of an angle is 6 times the complement. Find the angle.

8. The surface area of a cube is the same as the numerical volume of that cube. The edge of the cube is what?

9. 9.1235781235781. . . . After the decimal point, the 89th digit is what?

10. How many $7 tickets must be sold with $4 tickets so that the average of 24 tickets is $6?

Stop, oh please stop, and let's check the answers since it is the end of the test.

ANSWER SHEET

Part 1	Part 2	Part 3
1. _____	1. _____	1. _____
2. _____	2. _____	2. _____
3. _____	3. _____	3. _____
4. _____	4. _____	4. _____
5. _____	5. _____	5. _____
6. _____	6. _____	6. _____
7. _____	7. _____	7. _____
8. _____	8. _____	8. _____
9. _____	9. _____	9. _____
10. _____	10. _____	10. _____
11. _____	11. _____	
12. _____	12. _____	
13. _____	13. _____	
14. _____	14. _____	
15. _____	15. _____	
16. _____	16. _____	
17. _____	17. _____	
18. _____	18. _____	
19. _____	19. _____	
20. _____	20. _____	
21. _____	21. _____	
22. _____	22. _____	
23. _____	23. _____	
24. _____	24. _____	
25. _____	25. _____	

CHAPTER 26

PRACTICE TEST III
ANSWERS

PART I

1. x and y = ±3. You can't tell which one.　　　D

2. .5% = .005. 5/100 = .05. Biggggerrrr.　　　B

3. 1 yd = 3 ft. 6 ft × 9 ft × $10 = $540.　　　A

4. $(x - 3)(x + 2) = 0$. x = 3, −2 < 4.　　　B

5. x/.08 is greater than 1. If you square it it becomes bigger, much bigger. x/8 is less than 1. Bigger than 1 is bigger than less than 1.　　　A

6. If a = b, then the two are equal. If, let's say, a = 10 and b = 1, $a^2 - b^2$ is bigger. If a = 1 and b = 10, $(a - b)^2$ is bigger.　　　D

7. .26x = 88. Soooooo .52x = 2(88) = 176 is bigger than y, since .52y is only 44.　　　A

8. If angle C = 90°; $a^2 + b^2 = c^2$.
   If angle C is bigger than 90°, then side c has
   to be bigger and $a^2 + b^2 < c^2$.   B

9. 1/a is between 0 and −1, but if $-10 \le a \le -7$,
   thennnn 2a is between −20 and −14, much
   smaller.   A

10. Cancel the a + b from both sides and the
    question is which is bigger, −x or x? Since
    x is negative, −x is bigger.   A

11., 12. The most important fact is that a, b, c,
    are positive at least for 11. For 11, ab and bc,
    cancelling the b's we see a < c. So ab < bc.   B

12. We could have written a < b, b < c. If you take
    a number less and a number more, you can
    never tell which is bigger, under any
    condition.   D

13. Zero is the most troublesome number. 0/a = 0.
    0(a) = 0.   0 − 0 − 0 = 0.   C

14. One of the trickiest. Column A = $a^3/64$.
    Column B is $a^3/4$. You might think that B is
    bigger, buuuut not for negatives.   D

15. The larger side lies opposite the larger angle.
    n must be positive (it is on the side of a
    triangle). 8n > 7n.   A

16. Hopefully by looking, you can see that if you
    substitute x = 6, the two sides are equal. The
    2s at the bottom are there to throw you.
    $[3^2 = (-3)^2]$ D

17. Almost all the time it is better to simplify the inside of a parenthesis first. When you divide you subtract exponents.

$$\left(\frac{b^3}{a^2}\right)^3 = \frac{b^9}{a^6}$$

since a power to a power means to multiply exponents   A

18. The easiest way, I think . . .

$$\frac{6 \times 6 \times 6}{18} \, y. \text{ Now cancel. } \frac{\overset{1}{\cancel{6}} \times \overset{2}{\cancel{6}} \times 6}{\underset{\overset{\cancel{3}}{}{1}}{\cancel{18}}} \, y = 12y = 6.$$

Soooo y = 6/12 = ½.   B

Hopefully you did it a lot faster than it took me to type it.

19. Split it. $\dfrac{80 + 2x}{8} = \dfrac{80}{8} + \dfrac{2x}{8} = 10 + \dfrac{x}{4}$ E

20. Cross multiply. $\dfrac{x^2 + 9x + 8}{x^2 + 4x + 4} = \dfrac{1}{1}$

    $x^2 + 9x + 8 = x^2 + 4x + 4$. Cancel the x^2's.
    $9x + 8 = 4x + 4$. $5x = -4$. $x = -4/5$.   B

21. There are several ways to do this.

    a. PB = 6. BC = $6\sqrt{2}$, a 45-45-90 triangle.
       A = $(6\sqrt{2})^2$ = 72.   C

    b. PC is also 6. The area of triangle BPC is
       ½6 × 6 = 18.

There are 4 triangles 4 × 18 = 72. There are other ways, but all of them better give the same answer. Otherwise, there is no mathematics.

22. Take any number, say 5. $5^2 = 25$.
$(5 + 3)(5 - 3) = 16$. $16 - 25 = -9$. Orrrr
$(x - 3)(x + 3) - x^2 = x^2 - 9 - x^2 = -9$.
The answer is still A.   A

23. This is a toughie, because there are several
things you could do, but the correct one is to
just multiply it out.

$$\left(x - \frac{1}{x}\right)^2 = \left(x - \frac{1}{x}\right)\left(x - \frac{1}{x}\right) = x^2 - 2(x)\left(\frac{1}{x}\right) + \frac{1}{x^2}$$

$$= x^2 + \frac{1}{x^2} - 2 = 100.$$

Sooooo $x^2 + \dfrac{1}{x^2} = 100 + 2 = 102$.   D

24. A. 4 to 8 is a 100% increase. A is wrong.

    B. 6 to 9 is a 50% increase. 3/6 (try to leave
    as a fraction when you do these problems).
    Buuut 9 to 6 is –3/9, which is not a 50%.
    B is wrong.

    C. 8 to 12 is a 50% increase. 4/8. 12 to 6 is a
    50% decrease. 6/12. Answer is   C

25. This is not, not, not a ratio portion. The
question is how long between dings? When
you count, you start after the first ring. If a
bell rings 6 times in 5 seconds, there is one
second between rings. See the picture. So in
30 seconds, the bell rings 30 times + 1 at the
beginning, 31.   B

1 sec	1 sec	1 sec	1 sec	1 sec	
1	2	3	4	5	6 rings

PART 2

1. $4x + x = 180$. $5x = 180$. $x = 36$. $4x = 4(36) = 144 = y$. E

2. Remember, you are solving for x^4. Don't do toooo much. $x^4 = 64 + 17 = 81$. That's it! A

3. $(10ab^2)(10ab^2)(10ab^2) = 1000a^3b^6$. E

4. $\dfrac{4 \text{ zoups}}{11 \text{ zounds}} = \dfrac{7 \text{ zims}}{8 \text{ zims}}$ zims cancel

   $\dfrac{4 \text{ zoups}}{11 \text{ zounds}} = \dfrac{7}{8}$

   Multiply both sides by 11/4.

   $\dfrac{\text{zoups}}{\text{zounds}} = \dfrac{7}{8} \times \dfrac{11}{4} = \dfrac{77}{32}$ A

5. Quarters are 25¢. Dimes are 10¢. The total value is $10d + 25q$. E

6. x is the complement of 32. $x = 58$. $58 + 90 = 148$. E

7. 3 divided by 4/3m. Flip 4/3m upside down and multiply.

   $\dfrac{3}{1} \times \dfrac{3m}{4} = \dfrac{9m}{4}$ E

8. $(10 \times 10)^2 + 10(10)^2 = 10{,}000 + 1{,}000 = 11{,}000$. B

9., 10. $4 * 5 = 2(4)(5) + 5 = 40 + 5 = 45$. B

   $2ab + b = 0$. $2ab = -b$. $a = -b/2b = -\tfrac{1}{2}$. B

11. You must substitute at least 3 numbers to be sure, unless you know that the chart is the equation of a line. Then only 2 numbers are needed. The answer is $y = 3x - 2$.
$y = 3(1) - 2 = 1; y = 3(2) - 2 = 4$.
$y = 3(3) - 2 = 7$.   C

12. By trial and error $k = 30$. $30/2$ is odd and $30/5$ is even.   C

13. $x = \dfrac{ay}{b} + c.$ Multiply by b.

   $bx = ay + bc.$ Subtract bc.

   $bx - bc = ay.$ Divide by a.

   $\dfrac{bx - bc}{a} = y.$ E

Remember if you are terrible at problems like this, don't worry. If there are 8 problems you absolutely can't do, getting the other 52 right means 700 plus. You must concentrate on getting right what you know.

14. The area of a "ring" is the area of the outer circle minus the area of the inner.
$r_{out} = 5 + 2 = 7$. $r_{in} = 5$. $49\pi - 25\pi = 24\pi$.   C

15. Normally on a real SAT, there would not be two questions involving the area of a circle. But this is practice. The shaded portion issss the area of ¼ of a circle minus the area of a triangle.

   $A = \frac{1}{4}\pi 10^2 - \frac{1}{2}(10)(10) = 25\pi - 50 = 25(\pi - 2)$ B

16. You can substitute points, but the easiest way is to solve for x and the coefficient of x is the slope.

$4x + 5y = 20$ $5y = -4x + 20$

$$\frac{5y}{5} = \frac{-4x}{5} + \frac{20}{5} \qquad -4/5 \qquad\qquad\qquad D$$

17. $10^2 - (-8)(-8) = 100 - 64 = 36.$ A

18. The exterior angle is the sum of the remote interior angles. Angle A plus angle B = 150.
$3x + 2x = 150.$ $5x = 150.$ $x = 30.$ $3x = 90.$ D

19. $y = \dfrac{x^2}{z^3}$ $\dfrac{xy}{z} = \dfrac{x^2(x)}{z^3(z)} = \dfrac{x^3}{z^4}$ A

20. $9^{3x+3} = 27^{4x-5}.$ We must see that 9 and 27 are powers of 3. $9 = 3^2.$ $27 = 3^3.$

$(3^2)^{3x+3} = (3^3)^{4x-5}.$ If the bases are the same, the exponents must be the same.

$2(3x + 3) = 3(4x - 5).$

    $6x + 6 = 12x - 15.$

      $-6x = -21.$

        $x = -21/-6 = 7/2.$ C

21. This is an average problem, but if you add all of them up, you may take forever. The average must be 7 feet.

$6'7'' = -5''$ from $7'$.

$6'9'' = -3''$

$6'10'' = -2''$

$7'1'' = +1''$ Total $-9''$. So the center (the 5th player) must be $7'9''$, BIG!!!   C

22.

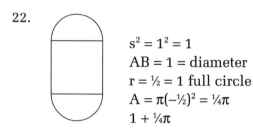

$s^2 = 1^2 = 1$

$AB = 1 = \text{diameter}$

$r = \frac{1}{2} = 1 \text{ full circle}$

$A = \pi(-\frac{1}{2})^2 = \frac{1}{4}\pi$

$1 + \frac{1}{4}\pi$ C

23. A of circle $= \pi(6)^2 = 36\pi$.

$\dfrac{24\pi}{36\pi} = \dfrac{2}{3}$ of ⊙

2 vertical ∡s at 0   1/3 of ⊙
$\frac{1}{3}(360) = 120°$.

1 angle at 0 $= \frac{1}{2}(120) = 60$.
$x = 90 - 60 = 30$.   B

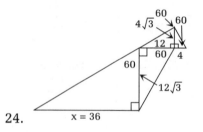

24. $x = 36$ E

25. Forget about last 3 zeros (32,~~000~~, 80,~~000~~).
     $.40(32) = 12.8$ $.15(80) = 12$.   A

The rest are totally different.

PART 3

1. After 2, all primes are odd. 51 is divisible by 3, as
    is 57. 55 is divisible by 5. $53 + 59 = 112$.

2. Trial and error. $6 \times 8 = 48$. $8/6 = 4/3$.

3. $\dfrac{7}{9} = \dfrac{56}{72}$ $\dfrac{7}{8} = \dfrac{63}{72}$

    57/72, 58/72, 59/72, 60/72, 61/72, 62/72 all OK.

4. $-7 + -6 + -5 + -4 + -3 + -2 + -1 + 0 + 1 + 2 + 3 + 4$
    $+ 5 + 6 + 7$

    $(-7)(-6) = 42.$

5. $4^{7+1} = 4^3 = 64.$ $n + 1 = 3.$ $n = 2.$ $2^{2+2} + 3^{2+1} + 5^{2-1} =$
    $16 + 27 + 5 = 48.$

6. $m = \dfrac{2a - 0}{0 - (-6a)} = \dfrac{2a}{6a} = \dfrac{1}{3}$

7. A toughie. x = angle. $180 - x$ = supplement of x.
    $90 - x$ = complement of x.

    $180 - x = 6(90 - x)$

    $180 - x = 540 - 6x$

    $5x = 360$

    $x = 72$

8. $x^3 = 6x^2.$ $x = 6.$

9. Don't look at 9. Repeats every 6 places. $6\overline{)89}$ $\quad^{14r5}$.
    Fifth place is 7.

10. Story and word problem. But, again, only worry
    about what you don't know. This really isn't too
    bad.

Cost/ticket	× tickets	= $
$7	x	= 7x
4	24 − x	= 4(24 − x)
6	24	= 144

$7x + 4(24 - x) = 144$

$3x + 96 = 144$

$3x = 48$

$x = 16$

**Trial and error might
work if you have the time.**

HOW TO TAKE THE REAL SAT®

You might ask, "Now that I've finished your book, what should I do?" If I were you, this is what I'd do.

1. Buy the ETS book, 10 SATs. These are real SATs to practice, which you want to do. There is extra stuff in the book that might help, but practicing real SATs is important. Practice them for time also.

2. Study as much as you want until two days before the SAT. Say our SAT is on a Wednesday, which it probably won't be. The last time I would study heavily would be Monday afternoon. After that, relax. I believe that your subconscious has answers. If you are relaxed, your subconscious may give you 50 or more points.

3. If you must study a little the last two days, only study a couple of problems, a couple of formulas, or a couple of words that will bother you if you don't know them. Last minute cramming on this test does not help.

4. The night before, have everything ready to go for the morning: test forms, entrance cards, ID, pencils, calculator (for security), and anything else you might need.

5. Go to sleep at the same time you normally do for school (unless you don't get enough sleep nor-

mally). If you go to bed too late or too early, you probably will not be rested properly.

6. In the morning eat breakfast, even if you do not normally eat breakfast. Experiments have shown students that eat, perform better.

7. Dress comfortably.

8. If you are a slow starter, that is it takes you a couple of minutes before you are going at full speed, do two easy math problems and two easy English problems. You must do the problems because all the tests are timed, and you must be going full blast immediately. They should be easy, so that you won't kill your confidence.

9. Bring some food and/or drinks with you to keep up your energy level.

10. Arrive at the room early. This will make you more relaxed.

11. It is OK to be a little nervous. Once the test starts you should be fine. The SAT is the one test in all of high school I wasn't nervous for. Regular high school tests, I would say to myself, "Did I study enough? Did I study the right thing?" However, no matter how much you study, the SAT can always put in something you've never seen. You'll probably miss it. There is nothing you can do. Just concentrate on getting what you know correct.

12. Guess if you don't know. If you are average, you will break even. If you are unlucky, you may lose a few more points. If you are lucky or you can eliminate one or more of the multiple choice answers, you may get a lot of extra points!!!!

13. Never spend too much time on one question.

14. Use your calculator as little as possible.

15. Never change an answer unless you are 100% sure. First answers are almost always best. Annnd

16. Have fun!!!!

GOOD LUCK!!!!GOOD LUCK!!!!GOOD LUCK!!!!GOOD LUCK!!!! GOOD LUCK!!!!

ACKNOWLEDGMENTS

I have many people to thank.

I would like to thank my wife Marlene, who makes life worth living.

I thank the two most wonderful children in the world, Sheryl and Eric, for being themselves. I would also like to thank Sheryl for helping to edit this book and Eric for the opening photo of me.

I would like to thank my brother Jerry for all his encouragement and for arranging to have my nonprofessional editions printed.

I would like to thank Bernice Rothstein of the City College of New York and Sy Solomon at Middlesex County Community College for allowing my books to be sold in their bookstores and for their kindness and encouragement.

I would like to thank Dr. Robert Urbanski, chairman of the math department at Middlesex, first for his encouragement and second for recommending my books to his students because the students found them valuable.

I thank Bill Summers of the CCNY audiovisual department for his help on this and other endeavors.

Next, I would like to thank the backbones of three schools, their secretaries: Hazel Spencer of Miami of Ohio, Libby Alam and Efua Tongé of the City College of New York, and Sharon Nelson of Rutgers.

I would like to thank Marty Levine of Market Source for first presenting my books to McGraw-Hill.

I would like to thank McGraw-Hill, especially John Carleo, John Aliano, David Beckwith, and Pat Koch.

I would like to thank Barbara Gilson, Mary Loebig Giles, and Meaghan McGovern of McGraw-Hill, and Christine Ducker and M. R. Carey of North Market Street Graphics, for beautifying this book.

I would also like to thank my parents, Lee and Cele, who saw the beginnings of these books but did not live to see their publication.

Last, I would like to thank three people who helped keep my spirits up when things looked very bleak: a great friend, Gary Pitkofsky; another terrific friend and fellow lecturer, David Schwinger; and my sharer of dreams, my cousin, Keith Ellis, who also did not live to see my books published.

ABOUT BOB MILLER...
IN HIS OWN WORDS

I received my B.S. and M.S. in math from Brooklyn Poly, now Polytechnic University. After my first class, which I taught as a substitute for a full professor, one student told another upon leaving the room that "at least we have someone that can teach the stuff." I was forever hooked on teaching. Since then I have taught at Westfield State College, Rutgers, and the City College of New York, where I've been for exactly 30 years. No matter how bad I feel, I always feel great when I teach. I am always delighted when students tell me they hated math before but now they like it and can do it. My main blessing is my expanding family. I have a fantastic wife, Marlene; a wonderful daughter, Sheryl; a terrific son, Eric; a fabulous son-in-law, Glenn; a marvelous daughter-in-law-to-be, Wanda; and my brilliant, adorable granddaughter, Kira Lynn, 16 months at the writing of this book. My hobbies are golf, bowling and doing crossword puzzles. Someday I hope a publisher will allow me to publish the ultimate high school text and the ultimate calculus text, as this brilliant publisher did by publishing my math SAT book so that our country will remain number one forever.

To me, teaching math is always a great joy. I hope I can give some of this joy to you.